1 MONTH OF
FREE
READING

at
www.ForgottenBooks.com

By purchasing this book you are eligible for one month membership to ForgottenBooks.com, giving you unlimited access to our entire collection of over 1,000,000 titles via our web site and mobile apps.

To claim your free month visit: www.forgottenbooks.com/free821039

ISBN 978-0-484-41434-0
PIBN 10821039

Popular
Science
MONTHLY

JAN. 1919
Volume 94-No. 1

*"THE WOOD THAT ALL ITS USERS PRIZE,
THE WISE INVESTOR SAFELY BUYS."*

A WORD

as to the

Reasons and Purposes

behind the

Cypress Pocket Library

Everybody likes to build, but nobody likes "repair jobs."

Repair jobs inevitably represent *an additional investment without
any addition to value.*

That point is worth digesting.

When you build, whatever you build, you like to build "for
keeps."

Some people change their minds about styles, in building the
same as in wearables; our tastes develop and result in changes in
our wants; but nobody changes his or her mind as to wishing to
get the greatest possible *endurance*, or *wear*, out of the things they
buy, and especially is this true of *building* investments.

Yet, singularly enough, so many people know so little about
woods and their relative values and special utilities; so many people
think that "lumber is lumber" and never attempt to *specify* the
KIND of wood they wish to use; so many people believe that repair
bills are "necessary evils,"that we believe we shall be able to render
a real public service by continuing the publication of THE CYPRESS
POCKET LIBRARY, convenient in size, authoritative in character,
of *probable* value as a technical guide, and careful and scrupulous
in its every statement or inference.

We have not, and do not, by any means, recommend the use
of Cypress without discrimination; Cypress is not the best wood for
every use; but where it IS appropriate it is so *emphatically* (and
demonstrably) the *one best wood* that the *many* should know
about it instead of the comparatively few who formerly profited by
their special knowledge.

WRITE FOR VOLUME I, with full text of U. S. Government
Report on Cypress, and containing complete list of all the 43 volumes in the library. Then write us for the volumes that will best
serve you.

It may be of interest that many of the volumes of The Cypress
Pocket Library have become established as standard works of
reference—text-books—in a number of eminent educational institutions and Governmental Departments. This is a gratifying tribute
to the broad and helpful spirit in which these booklets have been
produced, and more than justifies the theories behind the original
pioneer idea of such a Library for Lumber USERS.

*Let our "ALL-ROUND HELPS DEPARTMENT" help YOU. Our entire resources
are at your service with Reliable Counsel*

SOUTHERN CYPRESS MANUFACTURERS' ASSOCIATION
1240 Hibernia Bank Building, New Orleans, Louisiana,
or 1240 Heard National Bank Building, Jacksonville, Florida

**INSIST ON TRADE-MARKED CYPRESS AT YOUR LOCAL LUMBER DEALER'S
IF HE HASN'T IT, *LET US KNOW IMMEDIATELY***

CONTENTS

Modern Publishing Company
225 West Thirty-ninth St.   New York City

Popular Science
MONTHLY

225 W. 39 Street
New York City

CONTENTS

CONTENTS—*Continued*

FREE BE A
CERTIFICATED
ELECTRICIAN

YOU MEN LISTEN The country needs more trained, graduate electricians. Thousands have gone into the Government service and there is such an unusual demand for competent electrical men that I am making a wonderful offer at this time. HERE IS YOUR OPPORTUNITY! I want to send you my splendid offer now.

Don't hesitate because of age or experience. Young men, boys and old men must now fill the gaps and keep business going. DO YOUR PART. Prepare yourself for a real position, by my **Home Study Course** in *Practical* Electricity. I am Chief Engineer of the Chicago Engineering Works. I have trained thousands of men and can help you better than anybody else. We also have large splendidly-equipped shops where you can come at any time for special instruction without charge. No other correspondence school can give you this.

SPECIAL OFFER: Right now I am giving a big valuable surprise that I cannot explain here, to every student who answers this ad. Write today!

$46⁰⁰ to $100⁰⁰ a Week

Go after some real money. Qualify for one of the thousands of splendid positions open. All you need to start is a few months snappy, *practical* instruction from a competent engineer. Come to me—*NOW*. I'll give you my personal care to ensure rapid and efficient progress. My course is intensely practical. It is highly condensed, simplified, up-to-date and complete. I am so sure you will make a splendid success in this study, that I will *Guarantee Under Bond* to return to you every cent paid for tuition, if you are not entirely satisfied when you receive your Electrician's Certificate granted. vou as a graduate of my school.

FREE—Lessons and Outfit—FREE

Send me the Free Outfit Coupon at once. Do it *now!* For a limited period I am making a slashing cut in the cost of tuition, and giving each new student a grand outfit of Electrical Tools, Material and Instruments—in addition—*Absolutely Free*. I will also send you—free and fully prepaid—Proof Lessons to show you how easily you can be trained at home to enter this great profession, by means of my new, revised and original system of mail instruction.

ACT PROMPTLY

Get the full benefit of this great offer. Send the Coupon or a postal for free information without delay. Do it now—before my free offers and guarantee are withdrawn.

CHIEF ENGINEER COOKE

Chicago Engineering Works
DEPT. 31
441 Cass Street   CHICAGO, ILL.

Use this "Free Outfit" Coupon

CHIEF ENG. COOKE, Dept. 31
441 Cass St., CHICAGO, ILL.

Sir: Send at once—fully prepaid and entirely free—complete particulars of your great offer for this month.

Name.................:.................................

Address..

....

5

Here is Your Chance to Succeed
through ELECTRICITY

"$100 a Week, Nell!

Think What That Means to Us!"

"They've made me Superintendent—and doubled my salary! Now we can have the comforts and pleasures we've dreamed of—our own home, a maid for you, Nell, and no more worrying about the cost of living!

"The president called me in today and told me. He said he picked me for promotion three months ago when he learned I was studying at home with the International Correspondence Schools. Now my chance has come—and thanks to the I. C. S., I'm ready for it!"

Thousands of men now know the joy of happy, prosperous homes because they let the International Correspondence Schools prepare them in spare hours for bigger work and better pay. You will find them in offices, shops, stores, mills, mines, factories, on railroads, in the Army and Navy—everywhere.

Why don't *you* study some one thing and get ready for a real job, at a salary that will give *your* wife and children the things you would like them to have?

You can *do* it! Pick the position you want in the work you like best and the I. C. S. will prepare you for it right in your own home, in your spare time—you need not lose a day or a dollar from your present occupation.

Yes, you *can* do it! More than two million have done it in the last twenty-seven years. More than 100,000 are doing it right now. Without cost, without obligation, find out how you can join them. Mark and mail this coupon!

How it Feels to Earn $1000 a Week

By a Young Man Who Four Years Ago Drew a $25 a Week Salary. Tells How He Accomplished It

HOW does it feel to earn $1000 a week? How does it feel to have earned $200,000 in four years? How does it feel to be free from money worries? How does it feel to have everything one can want? These are questions I shall answer for the benefit of my reader out of my own personal experience. And I shall try to explain, simply and clearly the secret of what my friends call my phenomenal success.

Let me begin four years ago. At that time my wife and I and our two babies were living on my earnings of twenty-five dollars a week. We occupied a tiny flat, wore the cheapest entertainment—and dreamed sweet dreams of the time when I should be earning fifty dollars a week. That was the limit of my ambition. Indeed, it seemed to be the limit of my possibilities. For I was but an average man, without influential friends, without a liberal education, without a dominating personality, and without money.

With nothing to begin with, I have become the sole owner of a business which has paid me over $200,000 in clear profits during the past four years and which now pays me more than a thousand dollars a week. I did not gamble. I did not make my money in Wall Street. My business is not a war baby—on the contrary, many others in my line have failed since the war began.

In four years, the entire scheme of my life has changed. Instead of living in a two by four flat, we occupy our own home, built for us at a cost of over $60,000. We have three automobiles. Our children go to private schools. We have everything we want, and we want the best of everything. Instead of dreaming of fifty dollars a week I am dreaming in terms of a million dollars —with greater possibilities of my dream coming true than my former dream of earning fifty dollars a week.

What brought about this remarkable change? What transformed me, almost overnight, from a slow-going, easily-satisfied, average man—into a positive, quick-acting, determined individual who admits no defeat, who overcomes every obstacle, and who completely dominates every situation? It all began with a question my wife asked me one evening after reading an article in a magazine about a great engineer who was said to earn a $50,000 salary.

"How do you suppose it feels to earn $1000 a week?" she asked. And without thinking, I replied, "I haven't the slightest idea, my dear, so the only way to find out is to *earn it.*" We both laughed, and soon the question was apparently forgotten.

But that night, and for weeks afterward, the same question *and my reply* kept popping into my brain. I began to analyze the qualities of the successful men in our town. What is it that enables them to get everything they want? They are not better educated than I—indeed, some are far less intelligent. But they must have possessed some quality that I lacked. Perhaps it was their mental attitude: perhaps they look at things from an entirely different angle than I. Whatever it was, that "something" was the secret of their success. It was the one thing that placed them head and shoulders above me in money-earning ability. In all other ways we were the same.

Determined to find out what that vital spark of success was, I bought books on every subject that pertained to the mind. I followed one idea after another. But I didn't seem to get anywhere. Finally, when almost discouraged, I came across a copy of "Power of Will." Like a bolt out of a clear sky there flashed in my brain the secret I had been seeking. There was the real, fundamental principle of all success—Power of Will. There was the brain faculty I lacked, and which every successful man possesses.

"Power of Will" was written by Prof. Frank Channing Haddock, a scientist, whose name ranks with such leaders of thought as James Bergson and Royce. After twenty years of research and study, he had completed the most thorough and constructive study of will power ever made. I was astonished to read his statement that, "The will is just as susceptible of development as the muscles of the body." And Dr. Haddock had actually set down the very rules, lessons and exercises by which anyone

could develop the will, making it a bigger, stronger force each day, simply through an easy, progressive course of training.

It is almost needless to say that I at once began to practice the exercises formulated by Dr. Haddock. And I need not recount the extraordinary results that I obtained almost from the first day. Shortly after that, I took hold of a business that for twelve years had been losing money. I started with $300 of borrowed capital. During my first year I made $30,000. My second year paid me $50,000. My third year netted me $70,000. Last year, due to increased costs of materials, my profits were only $50,000, though my volume of business increased. New plans which I am forcing through will bring my profits for the present fiscal year up to $65,000.

Earning a thousand dollars a week makes me feel secure against want. It gives me the money with which to buy whatever will make my family happy. It enables me to take a chance on an investment that looks good, without worrying about losing the money. It frees my mind of financial worries. It has made me healthier, more contented, and keener minded. It is the greatest recipe I know for happiness.

Prof. Haddock's lessons, rules and exercises in will training have recently been compiled and published in book form by the Pelton Publishing Co., of Meriden, Conn. I am authorized to say that any reader who cares to examine the book may do so without sending any money in advance. In other words, if after five days' reading, you do not feel that the book is worth $3, the sum asked, return it and you will owe nothing. When you receive your copy for examination I suggest that you first read the articles on the law of great thinking; how to develop analytical power; how to perfectly concentrate on any subject; how to guard against errors in thought; how to drive from the mind unwelcome thoughts; how to develop fearlessness; how to use the mind in sickness; how to acquire a dominating personality.

Never before have business men and women needed this help so badly as in these trying times. Hundreds of real and imaginary obstacles confront us every day, and only those who are masters of themselves and who hold their heads up, will succeed. "Power of Will" as never before, is an absolute necessity—an investment in self-culture which no one can afford to deny himself."

Some few doubters will scoff at the idea of will power being the fountainhead of wealth, position and everything we are striving for. But the great mass of intelligent men and women will at least investigate for themselves by sending for the book at the publisher's risk. I am sure that any book that has done for me—and for thousands of others—what "Power of Will" has done—is well worth investigating. It is interesting to note that among the 250,000 owners of "Power of Will" are such prominent men as Supreme Court Justice Parker; Wu Ting Fang, Ex-U. S. Chinese Ambassador; Lieut.-Gov. McKelvie, of Nebraska; Assistant Postmaster-General Britt; General Manager Christeson, of Wells-Fargo Express Co.; E. St. Elmo Lewis; Governor Arthur Capper of Kansas, and thousands of others. In fact, today "Power of Will" is just as important, and as necessary to a man's or woman's equipment for success, as a dictionary. To try to succeed without Power of Will is like trying to do business without a telephone.

As your first step in will training, I suggest immediate action in this matter before you. It is not even necessary to write a letter. Use the form below, if you prefer, addressing it to the Pelton Publishing Company, 14-A Wilcox Block, Meriden, Conn., and the book will come by return mail. This one act may mean the turning point of your life, as it has meant to me and to so many others.

The cost of paper, printing and binding has almost doubled during the past three years, in spite of which "Power of Will" has not been increased in price. The publisher feels that so great a work should be kept as low-priced as possible, but in view of the enormous increase in the cost of every manufacturing item, the present edition will be the last sold at the present price. The next edition will cost more. I urge you to send in the coupon now.

When Your Heart's in Your Mouth-

Then is when Tire Chains prove their _Real_ value—they add so much to your brake power. Without them brakes would be useless.

It's these *unexpected emergencies* that make a driver think quick and act like lightning. *When suddenly* the children dash out from the pavement and are almost under your wheels before you realize it—you *instinctively* jam down your foot-brake and *frantically grab the emergency.*

What if your brakes slipped and didn't hold? Wouldn't the consequences be awful? It's positively criminal for a driver of a motor car to overlook even the slightest safety precaution. *Unquestionably the most effective supplementary addition to brake power* when the roads and pavements are wet and slippery, is in the use of

WEED TIRE CHAINS

Cars with *chainless tires* on wet–greasy–slippery pavements lack brake power to the same degree as they would if their brake linings were made of wet–greasy–slippery bands of rubber.

Wet rubber slips—never grips. It slides like a cake of soap on moistened hands. It lacks the bite and hang-on ability of chains.

Good brakes and Weed Tire Chains are undoubtedly the *greatest* factor in preventing motor accidents.

It's the height of folly to even attempt to drive without chains on all four tires when the roads are slippery and uncertain.

AMERICAN CHAIN COMPANY, INC.

BRIDGEPORT CONNECTICUT

In Canada: Dominion Chain Company, Limited, Niagara Falls, Ontario, Canada

Largest Chain Manufacturers in the World

The Complete Chain Line — all types, all sizes, all finishes—from plumbers' safety chain to ships' anchor chain.

Popular Science Monthly

Waldemar Kaempffert, *Editor*

January, 1919; Vol. 94, No. 1
Twenty Cents; Two Dollars a Year

Published in New York City at
225 West Thirty-ninth Street

"Dry Shooting" for Airplane Gunners

By Captain E. C. Crossman, U. S. A.

THE Yankee airplane machine-gunner has to lead the enemy plane with his aim, precisely as the duck-shooter has to get his line of fire out ahead of the whizzing teal to inflict hits. If the airplane gunner shoots straight at the plane going past, he hits where the plane was, but is not, merely because it requires time for the bullet to go even 200 yards, and the two planes are traveling at speeds never touched by wild fowl.

Wherefore the neophyte in the machine-gun game is taught first to make quick allowance, by means of his "ring sights," for a plane traveling past him on a level and at right angles to the line of fire, and then later to lead correctly for the plane traveling in various directions with relation to his line of fire.

Aiming Ahead of Target

The student is seated in a pivoted plane chassis which sways at the slightest movement. It is fitted with the service ring sight, by which auto-

© International Film

Shooting clay pigeons is easy compared with calculating speed and direction of this model of a German battle-plane

matic allowance is made to put the line of fire ahead of the enemy plane, while still aiming straight at the pilot of the other ship. By aiming through the side of the rear

ring, the gunner can still aim directly at the other pilot, but is really pointing the gun ahead of the other plane, just as one can do with a wind-gage sight on a military rifle.

Use of Miniature Airplanes

With the students seated behind dummy guns, the crews working the targets operate the miniature planes mounted on pedestals to simulate full-sized planes at a distance, traveling at various angles. The student becomes accustomed to laying the sight of his gun on the plane and making the allowance necessary to inflict hits.

The second stage is with actual machine-guns and ball cartridge. The guns are mounted to simulate machine-gun mounts on a plane, and the plane targets are operated by a hidden and protected crew, as in the first practice. The guns, however, do not hit the planes, which are stationary.

Seated in a pivoted plane chassis that sways to the slightest movement, the student aviator learns the first principles of sighting a machine-gun at a rapidly moving **enemy airplane**

© International Film

Practising with real machine-guns. The guns are mounted to simulate machine-gun mounts on an airplane, and the targets are operated by a hidden and protected crew
© International Film

They hit a target to the right or left of the plane, the change in position of the shot group being made, of course, by the necessary "lead" to hit a fast moving plane.

The distance from the plane to various spots on the target is known and charted out; so, with the dummy plane at a given distance and the pit screw, the position of the shot left shows immediately whether or not a real plane would have been hit—

whether the judgment and lead used was correct to hit.

More advanced courses include firing from a high-speed hydroplane or sea-sled at targets in the water, and firing from planes at objects on the water to show bullet strike.

How the Modern Grenadier Is Armed

An ancient practice that was revived by the trench fighting of the great war

THE hand grenade was used by the French as early as the sixteenth century. During the present war several types of this weapon have been evolved. There are two principal types of hand grenades in use. One type is exploded by a time-fuse, the other on striking the ground or some other resisting object.

The Besozzi grenade has a time-fuse which causes the explosion of the charge of the grenade five seconds after its ignition. The free end of the fuse has a match-tip which is ignited by striking it with a ring worn on the grenade-thrower's left hand. Within five seconds after ignition, the grenade must be at a safe distance from the thrower and within the enemy's lines. The cast-steel body of the weapon is deeply grooved to assist in fragmentation.

The impact type of grenade is represented by the pear-shaped grenade shown in the middle of the accompanying picture. It has a firing lever, with a cam which holds up the firing needle when the lever is folded against the side of the grenade. The end of the lever is held in place by a strap which must be removed before the bomb is thrown.

When the grenade leaves the hand

of the thrower, the arm of the lever is swung through an arc of 180 degrees by a spring, and the cam releases the firing pin.

The ball grenade is made to explode by a time-fuse. The fuse is enclosed in a wooden tube which forms the neck of the bomb. It is ignited by pulling out a friction pin. The soldier about to throw the grenade hooks the wrist strap of his throwing hand to the ring of the friction pin. When the grenade leaves his hand, the friction pin is jerked out, thereby igniting the fuse, which explodes the charge, usually after five seconds.

© Underwood & Underwood
The two principal types of modern hand grenades, those exploded by a time-fuse and those set off by the impact of striking the ground or some other resisting object

Will this Invention Increase the Ship Propeller's Efficiency?

TO increase the efficiency of the ship's propeller, it is suggested by an inventor to place a smaller propeller, but with a steeper pitch, in front of the usual propeller, and on the same shaft. The smaller radius of the auxiliary screw compensates for the greater pitch, reducing the resistance to the screw.

The inventor of this double propeller claims that the auxiliary screw causes a better utilization of the engine's power. The small screw placed in front of the usual propeller acts upon the water near the shaft, where the larger screw exerts but little power. He contends that the addition of the smaller screw will not increase the burden placed upon the engine of the ship, and that, with the same motive force, it will add from 25 to 30 per cent. to the driving power of the propeller.

By placing a second propeller on the same shaft, an inventor expects to increase the speed of a ship without adding to the engine's burden

There is no doubt that the addition of another screw of greater pitch would increase the driving power of the propeller a great deal, but it is by no means clear how this can be produced without adding to the burden of the ship's engines.

© Publishers' Photo Service

Will That Shell Hit the Mark? It Depends on the Wind

IF you want to hit the target with a bullet, an arrow, or even a stone, you must allow for the wind. When arrows were used in warfare, the air currents were tested by experienced men, who determined the "windage" by throwing tufts of grass into the air and observing their drift.

But that is not accurate enough for the modern artilleryman. What good is a tuft when the range of fire is five, ten, twenty miles? For the purposes of modern artillery practice, a method for testing the air currents has been evolved that is thoroughly scientific and surprisingly accurate.

Small rubber balloons filled with hydrogen, and of a conspicuous red, are sent up, and their course is closely observed through the telescope of a transit instrument. The vertical rise of a balloon of that type under given conditions is known. The angle of deviation for different heights is ascertained at intervals of a few seconds. These data are carefully recorded by the observer, and from them the angle of correction is computed—that is, the angle between the true sighting line of the gun and the line in which the gun must be pointed to hit the target.

When the balloon ascends, its course is watched to learn the wind's velocity and direction

This mass of steel is to be forged into armor-plate of the best grade and used on battleships; it takes a week for such a mass to cool

A Little Ingot of Steel Weighing 320,000 Pounds

THE mass of steel shown above is known as a steel ingot, and its weight as it stands—or rather hangs suspended from a crane—is 320,000 pounds, or 160 tons. An ingot is the name given to the shape of the cold steel as it comes from the mold into which it was poured in a hot liquid condition when it left the steel-making furnace.

Sometimes these molds are long and round, sometimes they are square, sometimes they are octagonal or fluted, and sometimes they are of the shape shown—rectangular masses several feet wide and a foot or two or more thick.

It is the latter, similar to the one illustrated, which are later reheated and forged into the armor-plate, several inches thick, which is used to protect the sides of battleships. It takes a week or two for these large masses of metal to cool.

The ingot proper, or that portion which is finally used to be forged down to armor-plate or other finished forgings, is the rectangular part. The circular part, called the head, cools last, and presses down upon the lower main mass as this cools, filling up any part of this that becomes hollow.

An Aid to School-Teachers

THANKS to the ingenuity of an American inventor, the school-teacher may utilize the small boy's energy to good purpose by allowing him to clean his eraser automatically. The essential feature is an all-metal trough into which a galvanized wire screen fits, causing the dust to fall into the dust receptacle below every time the eraser is set down.

The eraser, is held out of the dust, as is the crayon. The screen can be removed and the dust cleaned out as often as is necessary. As the trough is all metal, it cannot be marred by mischievous pupils. It holds fast to the walls without warping. The holder, which may be easily installed in any school building, is especially suitable for fireproof structures.

Back-Yard Coal Mines

COAL shortage has no terrors for people throughout great areas of North Dakota, Montana, and Wyoming. They have coal mines in their back yards.

The largest deposits of coal in the United States are the lignite coal-beds of the Northwest, as yet practically unexploited. In western North Dakota alone the U. S. Geological Survey estimates there are 500,000,000,000 tons. The majority of farmers either have lignite on their own farms or dig their supply from a neighbor's coal vein.

Chicago's Memorial Arch to Her Dead Heroes

A MEMORIAL arch in Grant Park, Chicago, is Labor's tribute to that city's gold-star heroes. The arch was unveiled on Labor Day, when 200,000 organized workers marched beneath it and paraded through the downtown streets.

The arch is 80 feet high. It is topped by a huge golden star, and is illuminated at night by three searchlights in red, white, and blue. On the pillars of the arch are printed in gold letters the names of every battle in which American troops have fought. The names of Chicago soldiers and sailors who have died are placed upon the pylons of the arch. It was designed by A. N. Rebori.

To Get a Realistic Snow Effect

AN attractive snow-scene effect in a window can be produced with tufts of cotton strung on black thread. A striped black-and-white background makes an effective setting, and the effect is heightened by suspending the tufts and placing the threads at irregular intervals.

The scheme may be used not only in windows, but also in the decoration of banquet rooms, stages, and other places.

A Night's Offering to Kultur

THE picture below shows a number of good reasons why the Germans were anxious to make peace. In the beginning the Hun used aerial bombs like these to strafe London and Paris, in the belief that indiscriminate murder of civilians would hasten a German victory. But gradually the supremacy of the air was gained by the Allies, and when a collection of bombs such as these represented one night's rations for one bombing squadron, Germany was forced to take her own medicine. The Allied flyers chose military objects as their marks, but the air invasion of Germany was none the less effective in calling forth the national yell of "Kamerad."

British official photograph

British official photograph

An Envelope-Ejector for the Addressograph

A MARKED increase in the speed with which envelopes may be addressed by the hand addressograph is made possible by a new attachment which ejects the envelopes after they are addressed. A small guide shoots the envelopes in the envelope box as the arm is lifted, placing them in a neat pile in the desired order.

A Hillside Cellar

W. F. HOLT, who is developing a model ranch near Los Angeles, has constructed a combination cellar and milk-house that is proving a most efficient means of practical conservation. It is built in a hillside.

The two rooms are ceiled, walled, and floored with concrete. The heavy doors are divided into upper and lower sections, and the upper halves may be swung outward, leaving screened openings. This is done in the evening, and the interior is thoroughly chilled by the night air. In the morning they are closed. By this means, the temperature is kept so low that ice is unnecessary for the preservation of milk and butter.

Thumbs Down on the Pay-Roll

THE pen is not always mightier than the sword. There were many Indian troopers in the war who bayoneted beautifully, but who did not know how to write their names. This hindered them very little except when they signed the pay-roll. Here we see how they did it: with thumbprints. According to the law of thumbs, no two are alike.

Bolstering a BrokenTree-Top

WHEN the wind split the upper part of the trunk of a fine old poplar tree in Central Park, Los Angeles, it looked for a time as if the only remedy were to cut the tree down. But the city forester came to the rescue with a discarded flag-pole, which he cleverly utilized as a splint.

The pole was set securely in the ground at the base of the tree, and heavy iron cleats were bolted around the broken part of the tree. As soon as the bark has grown over the break the pole will be removed.

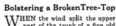

British official photograph

Germany's War on Pottery

THESE are not, as you might think at first glance, some new kind of mine, or great bombs, but merely a large collection of earthenware pots. They were originally manufactured for the arts of peace, and might have posed as models for the inanimate part of the scene in a modern conception of Omar's "A jug of wine, and thou, singing beside me in the wilderness."

But the Germans had to retreat, and, following their formula, "Smash everything you can't take with you," they destroyed the pottery factory, and sent the pots rolling over the fields, thus supplying the "wilderness" to go with the jugs.

Then came the British salvage corps, bent on saving all possible out of war's wreck, and took stock of the pots. The picture above shows members of this useful army division sorting the whole from the broken in an attempt to clean up the place which has now been made safe for democracy—and potters.

17

They Brave Death for a Picture

Desperate chances taken by the flying camera-men

THE life of an aviator in the British Royal Flying Corps is hazardous at all times; but there is one task that he often has to perform in which the danger incurred is perhaps the greatest of all. That task is the photographing of enemy positions from the air.

The pilot of a camera-plane must be a man who is not afraid to take any chance, no matter how desperate, in order to secure the desired photographs. His airplane is the finest of its type—generally a two-seater, equipped with a 160-horsepower engine. While not very speedy, this plane is easy to maneuver and very steady in the air. Three telephoto cameras are arranged so that they secure a triple panoramic view of the country below. The pictures are generally taken by the observer and not by the pilot. When he wants to make an exposure, the observer looks through a glass panel between his feet, which acts as a finder. The range of the cameras is really remarkable. I have seen photographs taken from a height of fifteen thousand feet which showed many details to the naked eye.

A Flight Over the German Lines

There are numerous thrills to be had on a picture-taking flight. If you will draw slightly on your imagination I will take you over the German lines with a camera-plane. Imagine that the aerodrome is somewhere on the British front in northern France, and that the camera-plane is already outside of its hangar. Attached to every squadron on active service in the Royal Flying Corps is a photographic section, which is supplied with a dark-room on a motor-

British official photograph · © Underwood & Underwood

Showing them how

Both single-seated and double-seated camera-planes are used. In the single-seater the camera is operated by the "joy-stick"; the plates are changed by the same means. A reflecting mirror constitutes the finder. Both planes are armed

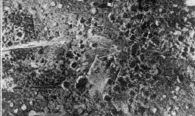

© Kadel & Herbert

Gaze upon the upper picture, and then upon the lower. Both are photographs taken from an airplane. They show the same village before and after bombardment

truck. This dark-room has all the necessary accessories for developing, printing, and enlarging, and is in charge of a sergeant and five men, two of them master photographers.

The cameras receive a final inspection by the sergeant. In the meantime, the pilot and observer are marking on their maps the area to be photographed. This is going to be a long flight, far into the enemy's territory, and the camera-plane's escort of ten fast fighters are being tuned up for the trip. You must remember that the camera-plane would be a wonderful prize for the Germans. It is the duty of these fighters to see that the precious plates are brought back safely. The weather conditions for the flight are ideal. The sky is absolutely clear.

Protected by Fighting-Planes

It does not take the observer long to decide on the altitude for taking the photographs. He knows, of course, that he has to fly high in order to avoid shrapnel and enemy planes on patrol. The cameras are pronounced "O.K." and orders arrive from G.H.Q. (General Headquarters) to proceed at once to the desired objective. The flight commander climbs into his fighting-plane. One after another, the escort of fighters float off into the air. The last to leave the earth, the camera-plane ascends slowly and takes its place in the "V" formation. The flight commander leads. Behind him are four more fighters. In the center of the V is the camera-plane, and bringing up the rear are the fighting pilots. Soon the planes are over No Man's Land and then over German territory. Don't think for a

18

The man at the left is not a pilot, but an observer. When he wants to make an exposure he looks between his feet at a glass panel that acts as a finder

When the prints are ready, a staff of experts reduce them to scale, determine where they overlap, and paste them together to form a photographic map

British Official Photograph

© Underwood & Underwood

moment that this group of machines flies level. As the shrapnel screams skyward, they commence to swoop and dodge, never for a moment, however, losing their compact formation. The pilot of the camera-plane sees by his instruments that he is up fourteen thousand feet; but even at this height the shrapnel from below tries to search him out.

Undisturbed, the planes wing their way toward their objective. It is cold at the height at which they are flying. A glance at his map tells the observer that he is almost over the territory to be photographed. Now he concentrates his attention entirely on the glass panel between his feet.

The Germans Try to Head Off the Camera-Plane

Far below are several rivers and canals, gleaming in the sun like tiny silver threads. The pilot is watching the flight commander's plane. As he looks the signal comes, as arranged, that the squadron has arrived. Throttling down his engine, the pilot puts the camera-plane into a quick-turning, rapid glide. Down it drops into the region where the shrapnel is bursting thick and fast. Up above, where the camera-plane has left them, the fighters circle around and around, protecting the camera-plane from attack.

Arriving at the proper level, and utterly oblivious to the shrapnel, the pilot straightens the plane out on an even keel for several seconds. As he does so the observer presses a button which unwinds the shutters of the cameras. As each photograph is taken, and before he pushes the handle that inserts new plates, he writes the number of the slide on the map of the territory which he has just snapped. Back and forth over its objective flies the camera-plane, until the observer is satisfied that he has photographed every inch of the ground below. Then a signal to the pilot, and the big plane ascends again to the level of its escort. Once more in formation, the fliers start on their way home.

Word of the evident success of the

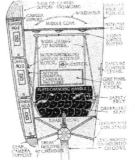

© Underwood & Underwood
As soon as the photographer touches ground again, he dispatches his camera to the developing-room installed on a motor-truck

Looking down on a camera-plane

raiders has been flashed to the nearest enemy aerodrome. The result is shown when a German squadron appears with the evident intention of cutting our squadron off. There is no question of

Don't think for a moment that photographing positions is as safe or as pleasant as yachting. The bullet hole tells why

dashing forward into the fight. The camera-plane, with its all-important information, must be seen safely home. The enemy squadron separates into groups of three, in an effort to lure one or two of our fighters to attack. But the formation must not be broken, and the tempting bait must be overlooked.

Nearer and nearer to the Allied lines flies the squadron, and the Germans, fearful lest their prey will escape them, swoop to the attack. Four of our planes immediately detach themselves from the convoy and engage the enemy, while the remainder of the convoy flies home with the camera-plane safely in its midst. The four planes left behind will keep the Germans busy. It is not long before the remainder of the squadron breaks formation and lands safely back at our own aerodrome.

Developing Photographs and Making Maps

As the camera-plane comes to earth, the sergeant and his assistants dismount the cameras and hurry away to the developing-room on the truck. Here the pictures are developed. Meanwhile the observer fills in a form, which is then sent to the sergeant to be forwarded to G.H.Q. with the finished prints. These prints, by the way, are finished in a little more than an hour after the plates are received. The moment they are ready, they are dispatched to G.H.Q., where a staff of experts reduce them to scale, determine where they overlap, and paste them together on the photographic map.

This great map has to be correct to the minutest detail. In order to keep it so, several small squadrons go out during the day, watching for changes over the enemy lines. The camera-plane of this small squadron is only a single-seater, equipped with one camera. An interesting feature of this plane is that the camera is operated from the "joy-stick" (control-lever) of the plane, even the changing of the plates is accomplished by the same means.

Yes, Santa Got Safely

And with Some Delightful

Dominos printed on cards are
easy for little hands to manage

In verse and picture this book
tells her the story of electricity

These lucky children have a modern toy town, with street-
cars, water-works, and "everything" run by electricity

Colored clay that doesn't need wet-
ting delights the budding sculptor

With the bath-
tub posing as
the ocean, chil-
dren get fun and
instruction from
this realistic
swimming toy

The new velocipede is strong
and like a bicycle in design, with
steel spokes and ball bearings

This thirteen-year-old is his own Santa
Claus: he has built a clockwork submarine that will actually run under water

This submarine "kiddy-car" carries a full magazine of happiness
and sinks nothing but dull care

He is learning word-building
with blocks that can be set
up on bases, sign-board fashion

When the rider mounts this
hobby-horse, a spring gives it
all the action of trot and gallop

through the War Zone

Excuses for Christmas Giving

This pistol fires coins, and a bull's-eye means another nickel in the bank

Cleanliness and art meet in this soap with ineffaceable pictures

A modern version of the old bean-bag game, with small shot-bags to release the beans for the table

A real player-piano in miniature delights the children—when the older folks will stop playing it long enough to give them a chance. The perforated rolls are run through the machine by turning a crank, and operate on the principle of the music-box

A nine-hole table golf course that is an improvement on the once familiar tiddledy-winks

Teaching the young idea to shoot with an unbreakable all-metal bow

There is endless fun for a boy in this boomerang device, which shoots a metal ring a considerable distance, and at the same time gives it a "reverse English" that causes it to return to the shooter

Should a submarine come down the street, this chaser will go into action at once

It needs the vivid imagination of children to detect the resemblance between this toy vehicle and an airplane

Going Motoring for Gold

COME now the automobile Argonauts, who make of treasure-hunting a picnic and pay the expenses of their holiday outings with the gold they glean so easily. This is in southern California, where "color" can be obtained in the sand of every mountain stream, and Sunday motoring parties have learned to take advantage of the opportunity to pick up pin-money and let the earth yield the cost of their tours. Within twenty-five miles of Los Angeles are the mouths of many great canyons, with trout brooks comparatively rich in placer, notably the San Gabriel, Santa Anita, Topanga, Pacoima, and Big Tejunga.

The expenses of their Sunday outing are paid with gold from the river

In the last, twenty-three miles from the city and within a hundred yards of an automobile drive, is a favorite spot where anyone who understands panning can readily wash out two or three dollars' worth in half a day, and stand a good chance of finding single nuggets worth five dollars or more; and here all summer Lewis W.

Within twenty-three miles of a city and just off the highway, she pans three dollars' worth of gold in half a day

Sometimes they take a tent along, and remain several days. Three members of the party do the panning, while a fourth takes care of the camp, which is supplied with fine trout from the creek. Mrs. Cubbison has become an expert at getting the yellow flakes out of the sand. On the whole, they have found the canyon so pleasant and profitable that Mr. Cubbison is considering permanent location there.

This place, like the others named, is in the Angeles National Forest, and open to everybody. The canyons have a large output of gold to their credit in years gone by, and big future strikes would not be surprising.

Cubbison and his family, who are residents of Los Angeles, have made almost weekly trips and more than defrayed their expenses with the proceeds. Sometimes the Cubbisons pick up burros, always available in these canyons, and make prospecting excursions far back in the high mountains.

Burbank Steps Forward with a Super-Wheat

AMERICAN wheat is one of our most important fighting assets. "Food will win the war." (You've heard the slogan before.) Of course, meats, fats, sugar, and leguminous foods, like beans, peas, and lentils, rich in muscle-building material, are also necessary; but nothing can take the place of wheat as a staple food. If grown in cold climates, it contains about 10 per cent of gluten, its most valuable constituent; if grown in hot climates, as much as 15 per cent.

Now comes Luther Burbank to the fore again. After experiments extending over eleven years, he assures us that he has evolved a "super-wheat" containing an unusually high percentage of gluten, and so sturdy that it may be grown anywhere from Labrador to Patagonia. Burbank's wheat resembles the prize-winning "Marquis," but has very large white, flinty kernels.

Luther Burbank says this wheat will grow anywhere; its food value is high

The new wheat is an exceedingly early grain—the earliest of some four or five hundred varieties which Burbank has been growing. He has tested it by comparison with sixty-eight of the best wheats of the world, and finds it superior in yield, uniformity, and all other desirable characteristics. It grows vigorously to a height of four feet on good ordinary soil, and thrives in almost any but the most extreme climates. On average valley soil, without special cultivation, care, or fertilizing, it produced in the past summer 49.88 bushels an acre.

Like all other wheats grown in California, the new wheat is a winter wheat. It was tried for baking bread, and the results of the tests were highly satisfactory. The loaves were of good color, texture, and taste. A high percentage of gluten in bread is of importance, because the food value of the bread depends upon it. Gluten has a high food value, and contains from 15 to 18 per cent of nitrogen in addition to carbon hydrogen, oxygen, and sulphur.

What Four Years Have Done for Flying

NOT one of the four types of airplane shown here was in existence at the outbreak of the present war. Army officers knew that the flying-machine would play some part in the war—but what part? No one divined that machine-guns would be fixed in position and fired through propellers; that speeds of 150 miles an hour would be about right for a fighter; that machines would be developed for special purposes—bombing, artillery control, fighting, etc. Four years of warfare have done more for the development of the flying-machine than ten years of peace-time discussion

23

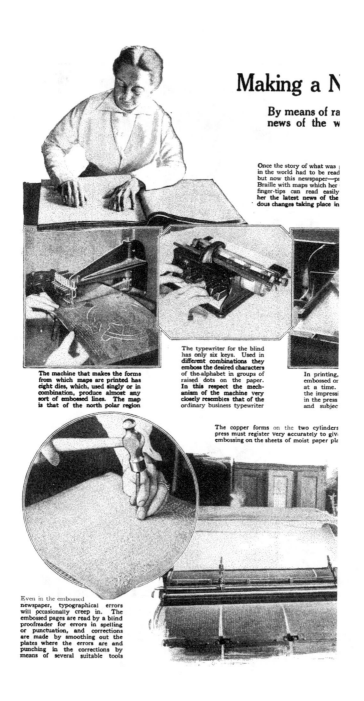

Making a N

By means of ra
news of the w

Once the story of what was
in the world had to be read
but now this newspaper—pi
Braille with maps which her
finger-tips can read easily
her the latest news of the
dous changes taking place in

The machine that makes the forms
from which maps are printed has
eight dies, which, used singly or in
combination, produce almost any
sort of embossed lines. The map
is that of the north polar region

The typewriter for the blind
has only six keys. Used in
different combinations they
emboss the desired characters
of the alphabet in groups of
raised dots on the paper.
In this respect the mech-
anism of the machine very
closely resembles that of the
ordinary business typewriter

In printing,
embossed or
at a time.
the impressi
in the press
and subjec

The copper forms on the two cylinders
press must register very accurately to giv
embossing on the sheets of moist paper pla

Even in the embossed
newspaper, typographical errors
will occasionally creep in. The
embossed pages are read by a blind
proofreader for errors in spelling
or punctuation, and corrections
are made by smoothing out the
plates where the errors are and
punching in the corrections by
means of several suitable tools

for Sightless Readers

embossed on manila paper, the
to the finger-tips of the blind

The revised Braille alphabet, which is
used in this newspaper for the blind,
consists of a combination of six dots
arranged in two parallel vertical rows.
The system also has special characters

Before the sheets of manila paper
are run through the press to be
embossed they are sprayed with
water. Because of the great strain
it must bear, only paper of a
very tough quality can be used

The sheets of which the newspaper
for the sightless is composed are assembled by blind workers. Each
pile contains copies of one sheet,
and the piles are arranged in
numerical order. The workers walk
around the table, taking one sheet
from each pile in proper order.
Each set of sheets is then fastened
together with wire staples by a
special machine before the newspaper is sent out to its blind readers

The dots of the
Braille alphabet are
punched into thin
copper plates by a
machine greatly re-
sembling a type-
setting machine.
One of the plates
serves as the stamp,
the other as the die
in the embossing.
They register per-
fectly, being made
by one operation

Here is a war map of the Balkan States and
of Asia Minor as it appeared in the newspaper for the blind. For obvious reasons,
the minor details of contour are omitted

When the embossed sheets come from
the printing-press they are still wet,
and they are dried on racks before
they are assembled in the final form

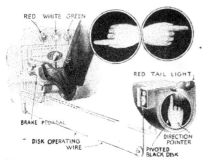

When the driver presses a foot-pedal, the hand points the way the car will turn

This is an apparatus for testing the efficiency of motor-driven automobile horns

For the Driver in the Car Behind

NOTHING except the movement of the driver's foot is required to operate the automobile direction signal recently invented by a Maine electrician and shown above. The device is automatic in action, being set in motion each time the car's foot-brake is applied. It consists of an illuminated rear-end signal carrying a white pointer hand on a black background; a special foot-brake pedal; a coil of wire attached to the brake mechanism to revolve a drum in the rear signal casing, operating the pointer hand and electro-magnetic buttons on each side of the pedal, which moves the pointer hand to the right or left to show the direction in which the car will turn.

When not in use, the hand is swung toward the front of the car out of sight from the rear. When the brake-pedal is depressed to slow down the car, the wire attached to the brake mechanism operates a small spring-driven drum to turn a shaft inside the signal casing, so that the hand becomes visible from the rear.

There is a switch to light the signal at night.

BECAUSE of the numerous accidents resulting to motorists by reason of failure to see embankments or curves in the road, the Automobile Club of Southern California is placing large reflecting danger signals throughout the State.

They show a beam of red light the instant the head-light of an approaching automobile falls upon them. The glass over the danger signal is clear around the edge and red in the center.

Picking the Longest-Winded Auto Horn

TYPICAL of the pains taken by automotive engineers to insure the most efficient design of even the smallest unit is an experiment made recently in selecting a motor-driven warning signal. Fourteen horns, two each of seven different makes, were packed together in a box filled with excelsior to deaden the noise.

Two large wooden pulleys were arranged so that one was driven by an electric motor and transmitted power to the second pulley by a belt. On one side of the second pulley was a strip of brass, semicircular in shape, which served to "make and break" the circuit that operated the horns.

The pulleys revolved at a rate that blew all the horns at once seven times a minute. Each time the circuit was closed for about four seconds, approximately the average length of warning signal the motorist gives: The horns were "tooted" ten hours a day.

The first to give out quit another, they became silent until the last one quit after sixty-eight hours, or a total of 28,560 "toots."

© Kadel and Herbert

Recently the Morristown and Erie, a small connecting road in New Jersey, has been trying out a forty-five-horsepower motor-truck with a carrying capacity of twenty-eight passengers, with good results

It usually takes two or more men to operate a so-called one-man automobile top; this can be managed by one

You do not need a magician's wand to turn your roadster into a touring car or a truck if it is equipped with this device

An Automobile Top Operated by the Engine

AN idea for a mechanically operated automobile top is shown above. It is operated by power secured from the car engine, and is based on the lazy-tong principle.

It consists of a vertical rack provided inside of the body at the extreme rear of the car. This rack is moved up vertically out of its compartment by means of two shafts and bevel gearing from power taken off the car engine through a meshing pinion set. The stop forces the lazy-tongs to extend forward until they reach the wind-shield side-bars, where they are attached, the tongs themselves forming the framework for the top material, which is folded up as the tongs are drawn close together before lowering the entire device into its compartment.

Roadster, Touring Car, and Truck, All in One

THE versatile folding-bed of the end of the nineteeth century is beaten to a frazzle by a recent invention in the automobile line. Charles J. Carlson, a citizen of Montana, has contributed an invention for changing an ordinary roadster into a touring car with baggage attachment, or into a truck.

When the car is to be used as a roadster, the rumble seat and truck body are folded up in the rear of the roadster seats. To convert the roadster into a truck, the rumble seat is left in its folded position, but hinged side pieces and an end piece are swung in position, so as to form a box for receiving the load. To convert the roadster or truck into a touring car, the rumble seat is unfolded on top of the box body.

Ingenious Coal-Gas Motor Tank

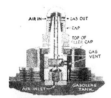

In these days of gasoline shortage, it is well to prevent the smallest leaks

IN Great Britain the development of the use of illuminating gas as fuel for motor-cars has been along two distinct lines: in flexible bag containers, and in comparatively small tanks capable of carrying the gas under small volume but at high pressure.

The tendency seems to be swinging toward the small, rigid tank. The tank here illustrated is light and yet small in size, and is capable of withstanding a pressure of from one to ten thousand pounds per square inch in a tank six inches in diameter. This is made possible by using two cast-steel ends and winding the tank in two layers of spirally wound steel tape in much the same manner as guns are wound. The longitudinal strength against rupture is secured by lengthwise wires of high tensile strength between the two ends of the tank, which are closed by steel-domed caps with expanding joints.

Seal the Gasoline-Tank

HAVE you ever looked at the filler cap of your automobile after a long trip through the country? If so, you undoubtedly found some moist dust around the cap. This was caused by an apparently small leakage out of the small vent-hole in the cap. Did you ever figure out how much gasoline was wasted by this seemingly small leakage?

Through tests conducted at the Armour Institute of Technology, a Chicago concern found that from five to eighteen per cent of the gasoline in your car is lost through evaporation and spillage, and it is now marketing a simple gasoline tank seal.

The seal permits the entrance of air into the gasoline-tank to replace the gasoline drawn out to the carburetor, but prevents the flow of gasoline out of the filler cap, either through spillage or by evaporation.

This coal-gas tank for motorcycles, etc., now being used in Great Britain, is the invention of S. J. Murphy, of Laurence Gate, Ireland

Dust is kept from the carburetor by this new type of air cleaner

A Dustless Carburetor

IN no part of an automobile or a tractor are surfaces so highly finished as they are in the engine. Between the piston and the cylinder wall a space no wider than the thickness of a hair is left—a space that is filled with a thin film of lubricating oil so as to produce a perfect seal. The efficiency of the engine depends in a large degree on the excellence of the fit between the piston and its cylinder. If the metal should be scratched there is bound to be trouble.

In a traction engine it is easy enough for dust to find its way into the carburetor and eventually into the cylinder.

An English firm has brought out a new type of air cleaner. Air is forced to pass through a water seal before it reaches the carburetor, and all dust is strained out.

Soldiers of the Saving Line

© International Film

An army may lay waste a country, but it wastes nothing itself. When it breaks camp, every scrap of material is gathered up. Here soldiers are collecting old wire, boxes, and oil-cans

Troops for overseas get new outfits. Their discarded equipment is carefully collected and sorted. Shirts, socks, and clothes are repaired and put to use again

© International Film

British official photograph

Even on the battlefield the salvage work goes on. No sooner has the attack been driven home than squads are sent out to collect the equipment of those who have fallen in the fight

The track of the retreating Germans was often marked by the wreckage of villages and homes. This salvage corps man, working within sound of the guns, is sorting and tagging the pathetic debris

© International Film

Currycombs, buckles, blacking-brushes, scrapers, and fly-killers were among the spoils left behind when the 76th Division went abroad. But not even an old tin can went to the rubbish heap—in army camps there isn't any such animal. Instead, down to the last fly-swatter, the cast-offs were picked up and listed by the station junk corps

British official photograph

This is an engine of war, but it belongs on the saving, not the firing line. It's a big straw-baler, and its business is to rebale straw on which potential heroes once slept—the heroes having gone to fight and the straw being needed for mattresses for the rookies

© International Film

28

Making Crippled Planes Fit Again

An airplane's flying life, even if it escapes accidents, is only about two weeks; therefore the repair shops are bustling places

Photographs © Underwood & Underwood

One of the store-rooms in which spare parts are kept ready for the demands of the airplane doctors

A hurry-up job on the fuselage. The repair corps' slogan is "Make 'em fit quick"

A single fast dive means taking apart the engine to clean out the soot

Testing an engine salvaged from a damaged plane. The engines have to be dismantled after every eighty hours of flying

A badly crippled airplane arriving at a repair depot. The engine alone is worth at least $5000. New machines are built from salvaged parts

This one-man tank, its inventor claims, can travel over land or sea, through sand or snow; the air is not included in his claims

The shell that smashed this tank cut a window opening for the Tommies who turned the monster into a snug hut

A One-Man Hand-Powered Tank

THE one-man tank above was recently patented by John Baptiste Felicetti, an Italian residing at Philadelphia. In outward appearance it resembles an Egyptian mummy-case on wheels. The shell of the tank, of bulletproof steel, conforms to the shape of the prone body of a man. On top is a hinged double door through which the soldier who is to occupy the tank enters. His body rests upon a hammock suspended from cross supports; his feet work the pedals of the steering-gear.

The mechanism for propelling the tank is rather primitive. On each side is a long lever with a handle on the front end and a spade-like pusher at the rear end. The man in the tank pushes the spade end into the ground, and propels the tank by pulling on the handles of the two levers. The armament consists of a machine-gun resting upon a swiveled support.

The inventor claims for his vehicle that it "can travel over land, sea, snow, and sand, or through water and mud." How did he overlook the air?

It Nipped Its Way through Wire Entanglements

THE combined length of the wire entanglements used in the war amounted to many thousands of miles. These wire protections proved to be highly effective as a weapon of defense, a sufficient reason to stimulate the ingenuity of inventors to devise some equally effective method of combating and overcoming this protection, without resorting to prolonged artillery fire.

Some of the earlier contrivances were so elaborate that they were unwieldy and an easy target for machine-guns and heavier artillery.

One of the simpler and more satisfactory devices, developed by the French, is a shield of steel, bent so as to protect the head of the wire-snipping soldier in front and on top. It is attached to the axle of two hollow wheels filled with sand to give them greater weight. The soldier using the device crawled toward the enemy's entanglement in the old Indian way, pushing the shield forward as he progressed. When he reached the entanglement, he thrust his wire nippers through an opening in the front of the shield and cut the wires.

The device is small, inconspicuous, and very effective.

How Are the Mighty Fallen!

ONE of the characteristics of an army is to make use of the very wrecks that war causes. A bulletpierced helmet becomes a decorative flower-pot in the back area quarters of some garden-loving soldier; a discarded canteen is the "makings" of a trench banjo; a wrecked automobile forms the groundwork of a Red Cross hut; and even the once mighty tank pictured here was turned into a cozy home for Tommies after a German shell had ended its fighting days.

This particular tank is one of the earlier types—great, heavy monsters that did wonders at first, but were too slow-moving to escape artillery fire; a direct hit usually put them out of action. The shattered side of this one, which makes a window for the improvised dug-out, is eloquent testimony as to what happened to tanks that couldn't dodge and twist through the enemy's fire.

The Tommies might be very happy in their bulletproof house, but the sight didn't please the men higher up. As a result, the "whippet" tanks were evolved—little fellows that could maneuver with all the agility of their namesake, and that didn't stay in one place long enough to give the artillery a chance at them. They quickly replaced the heavier early types for most work, and soon swarms of them were leading the attack and clearing the road for the infantry. Their efficiency helped to hasten the end of the war.

Photographs © International Film Service

Protected by this shield against rifle and machine-gun bullets, they cut barbed-wire fences

New Stars—What Are They?

A bright star suddenly flares up— why? Perhaps two stars collided

By Ernest A. Hodgson

THE stars, shining above us night after night, have changed little since mankind has lived on the earth. Many of the brighter ones bear names given them by the astronomers and astrologers of thousands of years ago. The constellations were arranged as they are to-day when the builders of the Pyramids observed them from those vantage points. So it is a rather startling phenomenon when a new star takes its place among the well known ancient configurations.

On June 8th last it was announced by Harvard that a new star had appeared in the constellation of Aquila. Harvard acts as the central bureau for sending out information about new discoveries to the astronomers of this continent. All new discoveries are reported to Harvard, and announced through its list of observatories. On the 7th of June there *was* a star in the position indicated, for it appears on some photographs taken then. Indeed, it appears on a photograph taken at Harvard in 1888 and in many others made since. But it was a star so faint that it could not be seen, except with a good-sized telescope.

It Seemed About to Outshine Sirius

On the 8th of June the new star blazed up until it was of the first magnitude. By the tenth it was almost as bright as Sirius, the Dog Star, and for a time promised to deprive it of the distinction of being the brightest star in the sky.

Sirius is about four and a quarter times as bright as a first-magnitude star. The title was quickly restored to Sirius, however; for in a few hours the new star was fading, and by the night of the 13th it was just first magnitude again.

There Are Many Wonder Stars

Since then it has continued to fade, with slight periods of recovery every ten days or so. By the middle of September it had become a good test of eyesight to find "Nova Aquilae," as it is called. By the time this article is published it will probably have become a telescopic star once more. But we shall not be able to tell; for by that time it will have passed, together with the other summer stars, and will be in the sky only during the day.

This Nova, although it is indeed a wonder star, is only one of many. Most Novae are discovered only by means of a painstaking study of star fields, photographed night after night. They are not noticed, for they do not

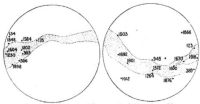

Most new stars appear in or near the Milky Way. Here we have plotted the principal new stars that have been observed and the dates when they were seen

become very bright—a large majority scarcely becoming visible to the naked eye at their maximum. Almost all the Novae, and practically all the visible ones, have appeared in that part of the sky known as the Milky Way.

Of the bright Novae, the one that has been studied most completely is that which appeared in the constellation of Perseus on February 22, 1901. This Nova became two and one half times as bright as a first-magnitude star within the space of a few hours. It then gradually faded, passing through many oscillations of about a magnitude, and finally dying away to a faint star once more.

Getting the Ranges of a Star

The distance of Nova Persel was finally computed. The method that astronomers use is, in theory, exactly the same as that followed by gunners who wish to measure the distance to a certain target. Their range-finder is provided with two prisms, which take the light from the two ends of a certain measured distance AB (page 32) in the instrument itself. The angle at A or B can be measured, and this angle, together with the distance AB, will enable them to find the distance OX. In the case of the range-finder, calculations are avoided by calibrating the instrument.

In determining the distance of a star, the line AB must be very long. The distance chosen for a base line was not a mile, or a thousand miles, or even the diameter of the earth; all these would be much too short. What they do use is the axis of the ellipse traversed by the earth in its journey around the sun—a base line 186,000,000 miles long. Observations are made six months apart, when the earth is first on one side and then on the other of its orbit. Even with this enormous base line, the angles, determining the distance of Nova Persei, were so small as to require great care and delicate instruments to measure. It was found that the Nova was so far away that light, traveling 186,000 miles a second, would require about 300 years to reach us from the star. So, although it was seen as a Nova in 1901, it really burst out into brightness about the year 1600.

Photographing the Stars

Probably the most interesting thing discovered about Nova Persei was discovered by means of photographs taken six months after it first appeared. These photographs were made by astronomers who carefully kept their telescopes pointing on the same point for hours at a stretch, following the star as it passed across our sky throughout the night. The light from the faint star was thus able to hammer away, so to speak, for all this time, impressing itself on the photographic plate.

The results showed that the Nova was apparently becoming surrounded with nebulous, cloudlike matter, which was passing farther and farther outward from the body of the star. Calculations showed that if this were being shot out from the Nova, it must be passing outward at about the speed of light. It has been suggested that the nebulous matter might have been there all the time, but that the light from the blazing up of the star had just reached it.

Sifting the Light of the Stars

If we look at the diagram we shall see what was meant. The light direct from the star reached us in about 300 years. That which went out to light up the cloudy material progressed outward for six months, traveling in that time about 3,000,000,000,000 miles. Then it was reflected toward the earth by the nebulous material, and took, of course, 300 years to reach us, arriving six months after the light direct from the outburst. And yet, so far away

was the Nova and its giant cloud that the whole appeared as a tiny speck, too faint to be seen except by a long exposure of the photographic plate, aided by the great light-gathering properties of a large telescope.

The photographs that appear on this page show what happens if we separate the various colors of light that come to us from the star, by means of a spectroscope,

TO THE EARTH 300 YEARS

NEBULA

TO THE EARTH 300 YEARS

The light direct from the new star (Nova Persei) reached us in about 300 years. That which went out to light up the cloudy material progressed outward for six months, traveling in that time about 3,000,000,000,000 miles. Then it was reflected toward the earth by the nebulous material, and took, of course, 300 years to reach us, arriving six months after the light from the first outburst. And yet, so far away was the Nova and its giant cloud that the whole appeared as a tiny speck too faint to be seen except by long exposure of the photographic plate, aided by the great light-gathering properties of a large telescope

How the distance of the earth from Nova Persei was computed

which is practically a light sieve. If all the light were, say, one shade of blue, there would be only a blue line showing. If all the shades of all the colors were present, we would call it white light, and the spectrum would be continuous from red to violet, as it is in the rainbow. The red light lies to the right-hand side of the photographs. These

spectra show that the star changed color from day to day. Dark lines indicate the absence of certain colors, and show that the substances which, in a state of incandescence within the star, would produce those colors are present in a cooler, gaseous state outside the star.

It will be noted that the spectrum of June 10 shows more dark lines than the one of June 9. This shows that the source of light was stronger, but was still shining through a gaseous envelope which was cool, compared with the center of the star. By July 30 the dark lines had almost disappeared and bright bands alone remained, indicating that the Nova was then heated throughout. The intensity of the light, however, was less, showing that the temperature as a whole by that time had fallen. This order of change in spectrum is characteristic of Novae in general.

Several Explanations

There are several explanations of the cause of these strange outbursts. One is that the star, as it traveled through space, ran into a mass of nebulous matter, which caused it to glow. This explanation, while it satisfies many of the observations, is difficult to accept; for the nebulous matter would need to be very dense to cause such an increase in tempera-

The spectrum of Nova Aquilae on June 9, 1918

The spectrum of Nova Aquilae on June 10, 1918

The spectrum of Nova Aquilae on July 30, 1918

These three photographs of the spectrum of Nova Aquilae, which were made by W. E. Harper, of the Dominion Astronomical Observatory, Ottawa, Canada, show how remarkably swift are the changes that take place in a new star. The straight lines and bands are produced by the glowing elements in the Nova. The same line or band always appears in the same relative place. Hence, as it disappears or becomes more intense we have unmistakable evidence of a great chemical change

ture, and it would need to be very small in area to account for the short time the star remained at its maximum. Another explanation is that the Nova collided with a dark star, the collision resulting in the enormous heat which caused the star to become gaseous and expand. It then cooled off gradually.

That New Dark Star

One aspect of this theory is interesting to us. May not a dark star even now be approaching, and about to collide with our sun, causing an outburst which would destroy the earth at once?

It may happen. Who can tell? But before it does a "new star" will appear in our sky—this dark star illuminated by our sun just as our moon is.

This star will *slowly* approach a maximum, however, and other phenomena connected with it

would enable astronomers to predict "the end of the world." We would have ample warning years before the collision.

But this collision theory does not explain the fact that the glow passes through other smaller maxima as it

dies away, and it has furthermore been calculated that there would have to be about four thousand times as many dark stars as there are bright ones to account for the number of Novae that have been seen—a rather difficult thing to believe.

This is a photograph of the nebulosity around Nova Persei, made by Yerkes Observatory. The light direct from the star reached us in about three hundred years. So, although it was seen as a Nova in 1901, it really burst into brightness about the year 1600

Absolutely "No Insurance"

So, the phenomena connected with Novae still await a theory that will combine them all. In the meantime, further evidence is being collected by our modern astronomical "adjusters" from our latest celestial "conflagration," which, it is hoped, will enable them to fix the responsibility. It may be safely stated, in any case, that everything on the dark body as well as on any dark satellite of the bright one that may have existed was a "total loss" and that there was absolutely "no insurance."

Transporting Huge Marine Boilers by Rail

STEAMSHIPS must have boilers, and these cannot always conveniently be built where the ships are constructed. Once safely stowed away in the hold of the ship, these gigantic boilers give little trouble; but to get them there safely is a different matter, especially if they have to be carried to their destination by rail. Some of these boilers have a diameter of 14½ feet and a length of more than 11 feet. Each weighs about 91,000 pounds, and a 3,500-ton ship requires two.

To transport these boilers on the regulation flat-cars is impossible, because, by reason of their large dimensions, they cannot pass through tunnels or under bridges.

The engineering problem presented in one case was solved by building a special car. The I-beams that form the longitudinal supports of the platform cars were curved in the middle

so as to bring them within a foot of the rails. In the cradle thus formed the huge boiler was placed, and securely fastened to supporting blocks.

When the steamer *Bear*, of the

This is one solution of the problem of shipping large marine boilers by rail

United States Geodetic Survey, was wrecked on the Pacific coast, many miles away from the nearest railroad line, great difficulty was experienced in transporting the huge boilers of the steamship after they had been salvaged.

The problem was solved by making them water-tight by plugging up all the openings, rolling them into the water, and towing the floating monsters to the nearest port. There the huge boilers were transferred to railroad cars to be hauled to their destination.

The proper making of these enormous boilers requires special machinery and appliances which are not always available at shipbuilding yards, and, after all, it is easier to transport a complete boiler than a whole factory with its equipment of specially designed machinery and its large force of trained workmen.

C International Film Service · C International Film Service

Those Wash-Days in Flanders

THIS husky standing by the machine is not a mechanic—he was a "washerwoman." His wash-tubs were the two big drums behind the engine, and in them he washed mud-, blood-, cootie-covered shirts and trousers for Yankee soldiers.

He would run his massive truck close up to the front lines, and dump the dirty uniforms into boiling water in the drums. Thus he kept the army clean and un-cootied.

The rifle on the front of his car was just a precaution in case the Germans couldn't resist the temptation to come over to steal the Monday wash.

"Here's Where We Licked Them"

MAPS are even more important to an army than to a stranger in New York's subway maze. Big war maps were in evidence everywhere on the western front. They were housed in shelters like the one shown above, where a British Tommy seems to be telling a squad of our boys just how and where it all happened. The map they are looking at shows the western front as a whole, but maps similarly sheltered show in great detail portions of the front, usually those directly opposite the trenches.

The quickest way to get maps is to send a photographer out with an aviator to fly over the country to be photographed.

How Did It Get Here?

IF you throw your wedding-ring into the ashes, do not be sure it is gone forever. It may come to life again, as did this one, around the neck of a potato. The ashes were thrown out on a field that had been tilled and cultivated for potatoes. One small, sentimental spud stuck his head through the ring, and when he grew up looked like this.

Youthful Aristocracy

THIS little aristocrat, who is not quite hardy enough to stand the cold winter wind, is comfortably fixed up in a limousine-sled. It has a heating and lighting apparatus within.

The small owner gets in through the top. When he is comfortably seated, his indulgent parent puts down the lid, picks up the ropes, and starts.

Turning Weeds into Sheep

A DISCOVERY that seems destined to play an important part in building up future supplies of wool and mutton has recently been announced by a Washington State sheep-rancher, Mr. Y. C. Mansfield. By mere chance he found that sheep thrive on the Australian salt-bush, heretofore regarded as a despised weed, of which there are literally millions of acres in some of our Northwestern States.

Motion Pictures Anywhere

AN enterprising manufacturer has developed a hand electric generator with which motion pictures up to eight or ten feet wide can be given.

The projector illustrated uses only slow-burning film, so that it can be used anywhere without the booth or insurance restriction. The entire outfit weighs only ninety-seven pounds. Probably the South Sea Islanders will soon be familiar with Mary Pickford.

British
official
photograph

A Rival to Houdini

IN far-off Bagdad, herculean strength was common in those "Arabian Nights" days when Aladdin wandered about with his magic lamp. The name "Bagdad" still holds its thrill for us, but British soldiers found the place disappointingly normal when they arrived there.

One day they happened on this Arab carrying on his back a life-boat—a task for ten ordinary men. Tommy, gazing at him, almost believed him to be a magic relic of the past.

When Paris Envied Pittsburgh

LIKE most other large cities, Paris has had its Society for the Abatement of the Smoke Nuisance: but circumstances alter cases. Such a small circumstance as one toot from the air-raid siren could convert every member of that society into an ardent worshiper of smoke; the thicker and oilier, the better they liked it.

The smoke-cloud defense, which proved so effective against submarines, was easily adapted to land, and, after trial at the front, was installed as a branch of Paris's air defense.

These smoke-producers, which freed a harmless smoke that was heavier than air or poison gas, were of modest mien. Objects that looked like concrete top-hats, inverted and scattered around the city in the path of any breeze, were the only apparatus used in the feat of blotting Paris from the face of France ten minutes after an alarm.

Danger !

WHERE three streets come together at one place the crossing is sure to be dangerous; particularly if traffic is heavy and there is a trolley line.

A startling danger signal has been invented for such a crossing. It stands on a corner that the cars pass, and is so connected with the track that when a car approaches the arm at the top revolves into an outstretched position and a large gong sounds. The signal itself is painted in startling colors, like a camouflaged ship.

When Huns Still Flew

WHEN airplanes flew low in support of infantry on the western front, there was no time for the anti-aircraft gunners to use range-finders. At any rate, the range-finders built in peace times were generally found to be useless against fast flying airplanes, and the successful aircraft hunter was he who had a quick and accurate eye for distance and speed.

But a machine-gun cannot be handled as easily as a 12-bore, and so an inventive Tommy hit upon the idea of mounting a discarded gun wheel on a post so that it would revolve, and then lashing his machine-gun to the rim of the wheel. This done, he was able to swing his piece in any direction skyward and make things warm for a low flying enemy aviator. But instances of machines being brought down by rifle or machine-gun fire were not common.

It Was Caught Before It Fell

HAD this pole fallen and the wires been injured, hundreds of telephones on the western front would have been put out of commission. Realizing the seriousness of this, these men worked desperately to keep it up.

A heavy snow-storm, followed by a freezing rain, weighed down the wires with frozen snow. The wires would have succeeded in dragging down the pole, owing to its insecure foothold in the soft dirt, had not these men interfered.

© International Film

British official photograph

Presenting the Bill to Bill

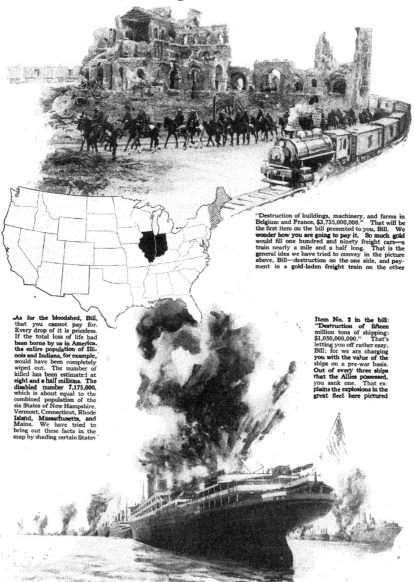

"Destruction of buildings, machinery, and farms in Belgium and France, $3,735,000,000." That will be the first item on the bill presented to you, Bill. We wonder how you are going to pay it. So much gold would fill one hundred and ninety freight cars—a train nearly a mile and a half long. That is the general idea we have tried to convey in the picture above, Bill—destruction on the one side, and payment in a gold-laden freight train on the other

As for the bloodshed, Bill, that you cannot pay for. Every drop of it is priceless. If the total loss of life had been borne by us in America, the entire population of Illinois and Indiana, for example, would have been completely wiped out. The number of killed has been estimated at eight and a half millions. The disabled number 7,175,000, which is about equal to the combined population of the six States of New Hampshire, Vermont, Connecticut, Rhode Island, Massachusetts, and Maine. We have tried to bring out these facts in the map by shading certain States

Item No. 2 in the bill: "Destruction of fifteen million tons of shipping: $1,050,000,000." That's letting you off rather easy, Bill; for we are charging you with the value of the ships on a pre-war basis. Out of every three ships that the Allies possessed, you sank one. That explains the explosions in the great fleet here pictured

The Pigeons' Flying Corps

These fearless aviators heed neither wounds nor exhaustion until they have reached home

This squadron's aerodrome was once a grocery wagon; later seventy-two flyers dwelt there in perfect **safety, for it was so well painted to match the landscape that it was invisible to enemy observers**

Photographs by Edwin Levick

Pencil and paper having been tied to the breast feathers of these airmen, they were ready for their day's work

With a pigeon in a cage on his back, ready to dash forth in search of his master, who is lost

This balloon is inflated in such a way that it will travel a limited distance and then gently drop to earth. A message is then written and attached to the pigeon balloonist, who when released flies swiftly home

If you whisper an address into a pigeon's ear, you can't expect him to go there; he is only taught to fly home. Send him by airplane, parachuting him down at the right spot

Make Your Own Christmas Tree

SANTA'S stinginess in handing out real pine Christmas trees in the last few years has caused us to look on him with increasing disfavor. But if our patron saint won't give us Christmas trees our inventors will make us some.

Here is a man-made tree that has several added attractions which Santa's kind doesn't have. Mounted on the base are a central shaft and an outside frame made up of poles which are held in place at the top by a piece of circular wood. These outside poles, which are stationary, have holes in them in which natural or artificial tree branches can be placed. The poles themselves incline inwardly, so when the branches are inserted this artificial tree will taper as a real tree does. On the central shaft are mounted several star-shaped platforms for holding angels, glittering balls, and any other decorations. This central shaft has a driving gear attached to it,

When the motor is started the star-shaped center revolves

The central shaft of this artificial tree, with its decorations, is revolved by a motor in its base

The mechanism is hidden beneath the store window floor   Clock machine

carefully concealed in the base. In the base there is also a motor with a pinion projecting upward which meshes with the gear. Thus when the motor is started the central shaft revolves, and so of course do the star-shaped platforms. The result is most pleasing. Through the spreading outside branches shine the revolving sparkling decorations within.

A storekeeper has fixed up a real live tree in his store window so that it will revolve. He trimmed off the lower branches of the tree, bored a hole through the floor of the window and inserted the end of the tree. On this he placed a split pulley and then fastened the end down to a revolving pedestal. He connected the pulley to a motor and started the motor. The motor made the pulley go round, and the pulley made the tree go round. And on the tree were many electric lights. And the results were great beauty and many sales for the enterprising storekeeper.

Showing how the motor turns the pulley and the pulley turns the tree

Counter stool

They're Ready for a Siberian Winter

FOR our troops in Siberia there is an enemy stronger and more relentless than the Bolsheviki; and that is winter. Napoleon's cohorts had a tilt with that same enemy in the less severe climate of Russia proper, and came off second best, finding on the great retreat from Moscow that the numbing cold clutching at their vitals and the drifting snow dragging at their legs were more deadly than the Cossacks.

In Siberia the Yankee fighting men will face a cold more extreme than Napoleon's troops knew. Even in the southern parts the winter bites deep. The extensive lowlands as well as the elevated plains lie open to the Arctic Ocean. The air, after being chilled on the plateaus, drifts down upon the lowlands, and in the region of the lower Lena extremely low temperatures obtain. Verkhoyansk is known as "the pole of cold of the eastern

atmosphere." The average temperature in December and January at Yakutsk is —40.2° F. and at Verkhoyansk —53.1° F., and occasionally the cold is so severe that official thermometers register as low as —75° and —85° F.

To work and fight in temperatures such as these, an army needs the equipment of an Arctic expedition. Fortunately, our own northern border is cold enough to have been a good laboratory in which to work out ideas for clothing soldiers in winter, and the Yankee in Siberia will be armed against the elements.

The War Department has devised fur-lined coats, fur helmets, fur gauntlets, and Arctic-proved footwear for our men. The equipment was assembled in San Francisco, and by the time you read this it has long been in service. A New York commuter looking at the pictures will sigh for Siberia and its comforts.

He is fully equipped with fur-lined parka, fur mittens and cap

With helmet closed and collar turned up, ready for a tilt with ice and snow

This Chilean desert was once the world's only source of fixed nitrogen

Nitrogen in War and in Peace

It helped to make the world safe for democracy and it keeps us from starving

By Frank Parker Stockbridge

NITROGEN, the most democratic of all the elements, is the essential factor without which the war for democracy could not have been won. In the last analysis, war is an effort to discover which of two sides can liberate the most nitrogen where it will do the greatest damage. For all explosives are nitrogen compounds—their deadly effects the result of the ineradicable tendency of this liberty-loving gas to burst its bonds and hurl in every direction fragments of whatever has served to restrain it. From old-fashioned black powder to the most modern and powerful trinitrotoluol, or "T.N.T.," nitrogen is the basis of them all.

Even more essential, in peace, is the possession of nitrogen in usable form. Without its aid there could be no plant growth, nor could animal life upon this globe continue. Yet, while nitrogen is literally as common as air, since four fifths of the volume of the earth's atmosphere is free nitrogen (serving to dilute the essential oxygen and make it breathable, so that you would not be literally burned alive), the problem of obtaining sufficient nitrogen is one that holds the serious, even apprehensive attention of scientists, economists, and the Government itself.

Before the world became densely populated with people who live in cities, and who therefore depend upon the annual crops produced by others for their food, instead of living on the

fruits, nuts, and game that were the food of our ancestors, nobody worried about nitrogen. People went where food could be obtained; if they failed to arrive soon enough they starved.

When the World Faced Starvation

Up to less than a hundred years ago, the entire human race was constantly menaced by the possibility of famine and wholesale starvation, and nature's methods of supplying nitrogen to plants through the action of bacteria in the soil long ago became too slow to keep pace with the increasing demand of the human race for food.

For a great many years the world has been dependent for its supply of nitro-

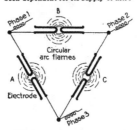

This shows the triangular arrangement of three Birkeland-Eyde single-phase furnaces with six adjustable electrodes

gen for fertilizer upon enormous deposits of sodium nitrate, or Chile saltpeter, found in the high, arid desert regions of Chile and Peru and nowhere else.

In late years there has been an important addition to this diminishing resource—the production of ammonium sulphate as a by-product of the coking of coal. But the total annual supply from both of these sources, about 2,500,000 tons from each, is still insufficient to meet the growing demand for agricultural purposes alone, while the war's demands created a situation little short of critical.

Crookes' Advice to Chemists

Twenty years ago, Sir William Crookes, then president of the British Association, startled us by declaring that the population of the world was increasing so much faster than its food supply that the race would soon face starvation unless new means of increasing the earth's fertility were found. His words carried conviction, and his suggestion that chemists turn their attention to the development of practical artificial means of extracting the nitrogen of the air and "fixing" it in usable compounds stimulated experiment in that direction.

As a laboratory experiment, the fixation of atmospheric nitrogen was old. The main essential of all processes then known, tremendously high temperatures, running up even to 6,000° F., made the practical application of any of them doubtful. Chemists, however, set to work. The development of hydro-electric power at Niagara and elsewhere, which made it possible to produce high temperatures through the electric arc, turned attention to this means of accomplishing the result sought. Charles S. Bradley, an American engineer, almost at the time Sir William Crookes was pointing out the imminent need, began the first large-scale experiment at Niagara Falls. His process was not a commercial success, but a little later a plant was established at Notodden, Norway, where water-power costs only $3 a year per horsepower. There nitrogen products were successfully made.

Reducing Nitrogen from the Air

The process used at Notodden and later at several other plants in Norway, known as the Birkeland-Eyde, is not unlike that devised by Bradley. An electric arc is produced by leading a current of about 5,000 volts equatorially between the poles of an electromagnet. This produces what is practically a disk of flame six and a half feet in diameter and having a temperature of about 3,000° F. The disk really consists of a series of successive arcs which increase in size until they burst. Air is passed through this arc.

The first product of the reaction is nitric oxide, which, on cooling with the residual gases, produces nitrogen peroxide. The cooled gases are then led into towers, where they meet a stream of water coming in the opposite direction. Thus nitric acid is formed

in the towers, in diminishing degrees of dilution. In the last tower the remaining gases are brought into contact with milk of lime, which combines with the gases to form calcium nitrate and nitrite. The nitric acid obtained in the other towers is combined with a base to form a commercial compound.

These Norwegian plants were financed by Germany, and their output of fixed nitrogen was almost entirely absorbed by that country. At the

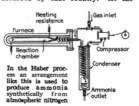

In the Haber process an arrangement like this is used to produce ammonia synthetically from atmospheric nitrogen

beginning of the war the annual production of the Norwegian plants was equal to about 10,000 tons of fixed nitrogen a year. At that time Germany imported annually 880,000 tons of nitrate of soda, equal to 137,000 tons of nitrogen. At the beginning of the war Germany had a stock of 1,000,000 tons of nitrate, equivalent to 156,000 tons of nitrogen, and a by-product ammonia capacity of 550,000 tons of sulphate of ammonia, equivalent to 113,000 tons of nitrogen.

It is known that the Norwegian plants have not been commercially successful. In Germany several other processes for the fixation of atmos-

pheric nitrogen were developed, all of which helped to supply the enormous quantity of nitrogen products required in manufacturing explosives.

One of these processes, developed by two German scientists, Drs. Frank and Caro, who began their experiments in 1898, is known as the cyanamid process. It is based upon the fact that calcium carbide, largely produced as a source of acetylene gas, may be induced with comparative ease to absorb nitrogen, thus forming a combination of calcium, carbon, and nitrogen, known commercially as cyanamid. This is the only process that has been installed on the American continent, a plant in Canada, at Niagara Falls, having been in operation for several years, with an annual capacity of about 60,000 tons.

Production of Cyanamid

The carbide is placed in the furnace and the reaction is initiated by local resistance heating to a temperature of 1500°-2000° F., the conversion proceeding to completion without further heating. The nitrogen is obtained from liquid air, manufactured by compressing air to a density of 500 pounds to the square inch and cooling by expansion. When the liquid air begins to rise above its normal temperature of —313° F., pure nitrogen boils off. The compound of calcium carbide and nitrogen, known commercially as cyanamid, is itself valuable as a fertilizer; by treatment with superheated steam its nitrogen may be released to enter into combination with the steam, forming ammonia. In Germany about 600,000

In the electric furnace shown here calcium carbide is made by fusing lime and coke at a high temperature. The carbide combines with nitrogen at 2000° F. to form cyanamid

Making pure nitrogen gas by distilling liquid air formed under a pressure of 500 pounds to the square inch at a temperature of 313° below zero, one of the features of the cyanamid process

tons of cyanamid is being produced annually.

The other process of fixing atmospheric nitrogen, on which Germany mainly relies, is the Haber process. In this, nitrogen and hydrogen gases, under a pressure of 1,500 pounds to the square inch, are passed through a chamber electrically heated to a temperature of 1,170° F. As a result, the nitrogen combines with the hydrogen to form ammonia. Although this process has long been in operation in Germany, its technical details have been carefully guarded, and it took chemical and electrical experts employed by the United States Government nearly a year to discover its secret from patents obtained in this country.

The Haber Process

The Haber process involves the presence of what is known in chemistry as a "catalyst." It has been found that certain elements or compounds —the list is constantly being enlarged—have the remarkable property of causing other substances to combine chemically, often in entirely new formations, without themselves undergoing any change or entering into the new combination. A familiar example is the common device for lighting gas without matches, which con-

sists of a small bit of "spongy platinum," or asbestos, fibers coated with platinum black. When this is held over an open unlighted gas-burner, the presence of the platinum causes the hydrogen in the gas to combine with the oxygen in the air with such speed and violence that great heat is generated by the reaction, the spongy platinum becomes incandescent, and in a second or two is so hot that the gas ignites.

In the Haber process the catalyst is spongy iron, although any one of several other substances probably would answer as well.

In the electric arc process the first product is nitric acid, which is directly usable for explosives. In the cyanamid and Haber processes the ammonia product is best adapted for use as fertilizer, but is readily convertible into nitric acid by passing a mixture of ammonia and air through a red-hot platinum screen acting as a catalyst.

The crude nitrate is shoveled into leaching-vats, where it is purified

The nitrate mining in Chile is nearly all surface work at little cost

The fact that nitrogen can be fixed directly in the form of sodium cyanide by the action of nitrogen gas on a mixture of soda and coke has been known for many years; but, while English, German, and American scientists have tried their hands on a commercial adaptation of this reaction, it is only recently that an American firm has been able to prepare sodium cyanide for the market by this process.

A Low-Temperature Process

Intensive study of the various methods of speeding up the reaction has led to the adoption of special apparatus, and, at a temperature around 1800° F., with the assistance of a specially developed catalyzer, an unusually pure cyanide is formed. From the cyanide it is easy to prepare ammonia quantitatively.

One advantage of this sodium cyanide method of fixation, aside from the low temperatures used, is that when ammonia is made from the cyanide, another product of commercial value is also obtained—sodium formate. This latter material can be used as a starting-point for a number of artificial flavoring oils, for a whole line of useful solvents, or for making the formic and oxalic acids that are so necessary in our dyeing processes; for instance, and which were formerly imported, chiefly from Germany.

Extracting Rheumatism through the Skin

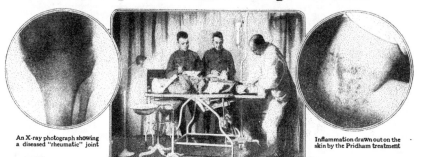

An X-ray photograph showing a diseased "rheumatic" joint

The soldier on the table is suffering with an ailment of a rib, which the surgeon is treating with Dr. Pridham's double nitro salts applied externally

Inflammation drawn out on the skin by the Pridham treatment

WHEN you have an ache in an elbow or a knee, you speak of your "rheumatism," and often attribute it to a change in the weather. But that same "rheumatism" is usually caused by germs, microbes, or bacteria located in some other part of your body.

Often there is a pocket of pus, or germs, hidden away in the root of a tooth you think is sound. One day you begin to have severe twinges of "rheumatism" in your big toe. That is because members of that little colony of germs located at the root of your tooth have started on the war-path. They are sending out a few scouts and much poison gas into your blood and lymph.

These enterprising little germs snoop around, keeping away from your leucocytes, or white blood corpuscles, and anti-poison tissues, always watchfully waiting for their big chance,

which comes when you inflict a slight injury on yourself. You may not notice that you stumbled and hurt that big toe joint, but the scouts from that germ colony noticed it, because they have found a weakened spot that cannot resist their onslaughts.

They arrive in droves and feed on that spot, lodging there until they have set up an inflammation which you call "rheumatism." Tubercle bacilli sloughed off the edge of a consumptive's lungs or intestines will do the same thing as the germs from the pus pocket on your tooth. Diseased tonsils, dysentery bacilli in the bowels, scarletina germs in the throat, pneumonia cocci in the lungs, adenoids, stone in the bladder, urological diseases, or any

spot where a nest of germs is located, may be the starting-point of "rheumatism."

Dr. Fred Pridham, of Johns Hopkins Hospital in Baltimore, has devised a treatment for joint affections that is being used very successfully at the military hospitals where it has been introduced. He uses a mineral, double nitro-peroxide, which actually extracts the inflammation.

The salt is mixed with a starchy substance which holds its action until it is placed on the skin over the diseased bone, from which it extracts the impurities, the germs, and their poisons.

The material is sprinkled on lint or gauze and placed on the skin over the diseased bone. When the doctor removes the dressings from six to twelve hours later, there will be seen blisters on the skin. If there is no infection the salt will not affect the skin.

Voice-Tubes for the Ship's Gunners

IN a naval battle, the range is obtained principally by men stationed in the mast-tops, the readings of the instruments being telephoned down to the officers in the plotting-room below the warship's deck. Here the instrument readings are quickly transcribed into terms of gun ranges and of angles of horizontal deflection.

These calculations are then sent to the gunners through speaking-tubes, although telephones and numeral indicators are often used, to make sure that the orders will be understood. For, when the battle waxes hottest, either a voice-tube or a telephone is likely to be swept away.

In big battles the gun that has but one channel of communication stands grave chances of being cut off from the

Here is a voice-tube equipped with a megaphone outlet which increases the intensity of the sound. This photograph was taken on board of the U. S. S. *Jouett*

rest of the ship. Should that happen, the gunners would have to depend upon the gun's telescopic sights, and there would be no checking up of hits or misses by the spotters in the mast-tops.

Thus the means of communication is the crux in the modern method of pointing and firing a battleship's guns. In our Navy, voice-tubes are generally preferred to electrical apparatus.

The speaking-tubes are metallic pipes made airtight to conduct the sounds efficiently. Nothing short of a rupture or a large hole torn in the tube by shellfire will impair its operation to a dangerous extent.

Press the Button—See the Show

How a little switchboard runs a theater
employing more than a thousand persons

By A. M. Jungmann

YOU are going to forget your cares and worries in an evening at New York's Hippodrome. You go early so as not to miss anything. Just as you settle into your seat the orchestra begins to play. In a few moments the show is on. The great red velvet curtains are drawn back, and tableau after tableau is presented to your eyes. Hundreds of performers throng the stage. Elephants and horses do their acts. The great tank is opened, and diving girls splash in.

Repairs Behind the Scenes

Between the first and second act, your wife turns to you with, "Oh! Look, Henry, that lovely velvet curtain has a big tear in it!"

You look, but don't see the rent. While your wife tries to show it to you, the

little electric switchboard operated by Clyde W. Powers, the stage director. Every time Mr. Powers pushes a button he starts something. When the orchestra begins to play, it is because

Mr. Powers communicates the need of a new light by pressing a button.

Each act is carefully timed. So many minutes must be allowed for a certain actor to do his part, so many seconds for another. Mr. Powers counts the minutes and the seconds as a miser counts his gold—nothing escapes him. In a sense, the audience is timed too, for favorite actors are permitted a few seconds beyond the time for their acts in order to receive the applause of the audience. After a show has been on a few days, the time the big hits require for applause is definitely set, and no actor may bow more bows than can be made in that time. When the last second arrives, one of the little buttons on the switchboard is pressed and the next cue is taken.

Suppose this efficient

How the show at the New York Hippodrome is directed. The stage director doesn't yell, "Here, you elephant man, bring us your elephants." He just presses a button. In one hundred and seventy-four minutes two hundred and twenty cues are given

Details of the wonder-working switchboard which controls the entrances and exits of 1074 persons

Should the switchboard fail (it never has) the telephone is at the service of the stage manager for the emergency

curtains are drawn back and the next act begins.

"Never mind, Henry," whispers friend wife, "I'll show it to you next time."

Womanlike she doesn't forget; but when the great red curtains are drawn across the stage again, your wife can't find the tear. You think it wasn't there. It was. While you were watching the stage, busy tailors neatly darned the rent.

Suddenly you say to your wife: "Say, there must be an awful racket out back of the scenes. Directors yelling at all that crowd and getting everybody in on time."

"I don't hear it," she replies.

Don't look patronizingly at her. There isn't any noise—no one yelling orders. The whole show is run by a

a button has been pressed. Everything you see done is done in response to a button on that little board. Actors, musicians, property-men, carpenters, electricians, tailors, animals, all obey the buttons. No one says anything. One thousand and seventy-four persons help make your evening a success by obeying the orders of these little electric buttons.

Even the Audience Is Timed

The performance lasts just one hundred and seventy-four minutes. In that time Mr. Powers gives two hundred and twenty cues by pressing the buttons on his switchboard. If an electric light bulb is not lighted, no one shouts for the electricians;

little switchboard were to get out of order, what then? There is a telephone, and, if that fails, a speaking-tube is ready to carry the orders that direct the show. But, so far, the little switchboard has never failed.

Obedience the Watch-Word

The same system is used in rehearsals. At no time is shouting or confusion permitted behind the scenes. So thoroughly is the idea of immediate obedience ingrained in all Hippodrome employees that a signal must be obeyed no matter when it is received. If the elephants were summoned on the stage at an hour when the elephant man knew there was no rehearsal, he would lose no time in marching his beasts up the runway and on the stage.

Before You Tried It On Your Piano

A catchy cover is almost as important as a catchy tune. Examining designs for a hoped-to-be song hit

One of the first steps is to photograph the artist's cover picture to make it ready for engraving

Printing a sheet on metal by a photographic process preparatory to etching the plates from which the song will be printed

At this stage the plate from which notes and words are to be printed is etched in an acid bath

The notes are very hard to etch; so, after the acid bath, they are hammered in by an expert

The "router" must carefully cut away the metal to make the notes and word type ready for printing

The song being fed to the binding machine, nearly ready to go out in its glory of fresh print to a waiting world

Cold? Press the Button

Carry an electric stove in your glove, and fight chills with the modern substitute for mustard

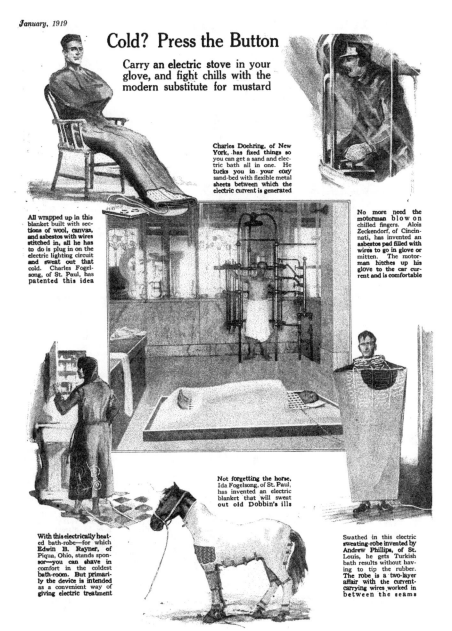

Charles Doehring, of New York, has fixed things so you can get a sand and electric bath all in one. He tucks you in your cozy sand-bed with flexible metal sheets between which the electric current is generated

All wrapped up in this blanket built with sections of wool, canvas, and asbestos with wires stitched in, all he has to do is plug in on the electric lighting circuit and sweat out that cold. Charles Fogelsong, of St. Paul, has patented this idea

No more need the motorman blow on chilled fingers. Alois Zeckendorf, of Cincinnati, has invented an asbestos pad filled with wires to go in glove or mitten. The motorman hitches up his glove to the car current and is comfortable

Not forgetting the horse, Ida Fogelsong, of St. Paul, has invented an electric blanket that will sweat out old Dobbin's ills

With this electrically heated bath-robe—for which Edwin B. Rayner, of Piqua, Ohio, stands sponsor—you can shave in comfort in the coldest bath-room. But primarily the device is intended as a convenient way of giving electric treatment

Swathed in this electric sweating-robe invented by Andrew Phillips, of St. Louis, he gets Turkish bath results without having to tip the rubber. The robe is a two-layer affair with the current-carrying wires worked in between the seams

Making Your Canary Feel at Home

If you take the proper care of your canary, you'll never need to ask, "Dicky bird, Dicky bird, why don't you sing?"

ABOUT fourteen distinct strains of canaries are known, and an almost endless number of varieties. The common canary is reared primarily for its song.

Before the war song canaries were reared in Germany; now most of the singing birds come from England. There is no reason why Americans should not train their own singing canaries. It seems to be necessary merely to put the birds in a room with good songsters, the vocal powers being developed through imitation. But careful watch must be kept, and a bird that develops harsh notes must be promptly removed; he must not corrupt the purity of the tone of his brothers. Sometimes a mechanical "thriller" is utilized in training, when the birds are silent, as during the moulting period. In six months or less the education of a singing bird is completed.

Mr. Alexander Wetmore, Assistant Biologist of the Biological Survey, U. S. Department of Agriculture, has given the subject of the management and care of canaries so much attention that we cannot do better than to abstract one of his papers on the subject.

Make the Bird Comfortable

First of all, he assures us that cages should be chosen primarily for comfort. The square cage, in his opinion,

The crest of feathers on its head sets this bird apart from the every-day canary

is best. For a single bird, the cage should be at least 9½ inches long, 6½ inches wide, and 9 inches high. A fine-mesh screen may be secured from the dealer and fastened around the lower half of the cage to prevent the scattering of seeds and seed-hulls. A common substitute for this is a muslin bag, held in place by a draw-string around the middle of the cage.

In a cage of the ordinary size three perches are sufficient. One should be placed at either end, at a distance that will allow easy access to the food and water receptacles, and the third elevated above the middle of the cage. For reasons of cleanliness, one perch should not be directly above another.

An English authority gives the standard size for

This bird, a crested canary, is like a Skye terrier inasmuch as its eyes are entirely concealed by its topknot

A Scotch fancy canary bird—bred to develop the hump on its back

Instead of being yellow, this bird, the lizard canary, has variegated plumage; hence its name

breeding cages as 22 inches long, 12 inches wide, and 16 inches high. Several types of open breeding cages made of wire may be obtained, or a box with a wire front may be made.

Care of the Cages

Though canaries, when acclimated, can endure a great degree of cold, they are very susceptible to sudden changes in temperature, and cold drafts are fatal. This should be borne in mind in choosing a place for the cage. The room must be kept at a fairly even temperature, day and night. Wherever placed, the cage must be kept scrupulously clean. Seed supplies must be replenished and water renewed each day. The receptacles for these necessities should be cleaned and washed carefully at short intervals. Cages that have removable bases should have the tray in the bottom covered with several thicknesses of paper. A better plan is to use the heavy coarse-grade sandpaper, known as gravel-paper, that may be secured from bird dealers. This should be renewed whenever the cage is cleaned, and in addition the pan should be washed in hot water from time to time.

What to Feed the Birds

The food requirements of canaries are very simple. The prime requisite is a supply of canary seed to which is added a small quantity of rape seed and a little hemp. If canaries do not seem to thrive, it is well to examine the seed supply and crack open a few of the seeds to make certain that empty husks alone are not being fed. Too much hemp seed should be avoided as it is very fattening.

In addition to this staple diet, lettuce, chickweed, or a bit of apple should be placed between the wires of the cage frequently. Bread that has been moistened in scalded milk, given cold, is also beneficial at times. Perhaps once a week egg food may be given. This is prepared by mincing an entire hard-boiled egg and adding to it an equal quantity of bread or unsalted cracker crumbs. Cuttle-bone should always be available to the canary.

In the breeding season egg food must be given daily as soon as the birds are paired. It may be discontinued or given at intervals of three or four days when the female is incubating. The yolk of hard-boiled egg only may be given for the first day after the young hatch. Seed should always be provided.

Under normal conditions most birds probably bathe daily. China dishes that are not too deep make good bathing receptacles. In winter the water should be warmed until tepid.

Another canary with a hump on its back, the Belgian fancy; it is a very rare bird

46

The Community House for Soldiers

An experiment in democracy conducted by several war organizations and financed by the people of Ohio

By Raymond B. Fosdick

AN army representative of the highest type of American democracy cannot be turned out by military machinery alone. We have come to regard the contentment of men in training as vital.

To the American Recreation Association, operating in their war work under the name of War Camp Community Service, fell a difficult task. Making sure that the men find available the best, and only the best, that the particular community affords, has been their task. Into more than six hundred towns and cities the W.C.C.S. has sent trained workers. In stimulating each community to utilize its own resources in the most effective manner, some unique institutions were evolved.

A Unique Community House

The first Community House was built at Manhattan, Kansas. The most interesting one, structurally at least, is at Chillicothe, Ohio, near Camp Sherman. It is in the form of a great Maltese cross. The people of the State of Ohio were asked to furnish the funds for its construction and equipment.

The Camp Sherman Community House has become the Mecca of people from all over the country who come to be near their sons, husbands, sweethearts, and brothers in the camp. Here, too, the townspeople mingle freely with the soldiers; and the community activities, the musical life of the town, home talent entertainments, and social life generally is conducted on a scale heretofore impossible with small-town facilities.

An Experiment in Democracy

The lounge, occupying the entire central space of the building, opens on one side into the restaurant and on the other into the auditorium with its stage extending across one end; and when these three rooms are thrown into one, as can be easily accomplished, their seating capacity is fifteen hundred. A large reading and writing room occupies the fourth wing of the house, and from it open women's rest rooms and small parlors.

Built on the curved driveway leading from this central building to the main thoroughfare, are the other hostelries of the "community"—each a model lodge in itself, harmonizing in color and architecture with the main structure. These provide sleeping-rooms with baths, writing-rooms, smoking-rooms, and women's parlors.

The Community House was an experiment in democracy. The results are gratifying. There is noticeable between the officers and men a cordiality that old-time military authorities would have declared impossible.

The change in the attitude of commanding officers in regard also to bringing women into the camp environment is interesting to observe.

A Contribution to Community Life

It was General Glenn who did most to hasten the building of the Camp Sherman House, and to make sure that every comfort was furnished the families of the men coming there. The Community House means the re-establishment of the normal social relations of life for the service men, and I am sure that every dollar which Manhattan, Kansas, or any other town has raised by bond issue, or the State of Ohio or the Rotary Club have pledged, will contribute not only to the soldiers' need, but to attaining the ideal town life.

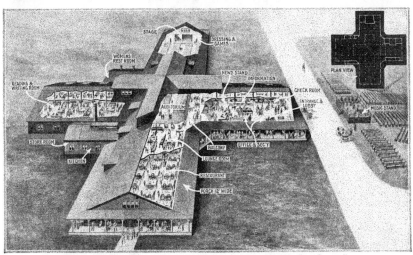

It was constructed jointly by several war organizations, and administered by War Camp Community Service, its object being to promote cordial social relationships between civilians and soldiers

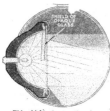

This shield of prismatic glass deflects the glare of your lamp

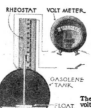

The rheostat indicates on your voltmeter the number of gallons of gasoline in your tank

Some New Answers

What invention and ingenuity
and comfort of truck,

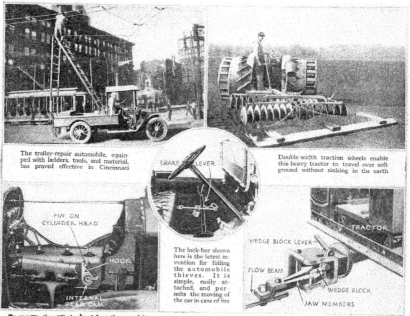

The trolley-repair automobile, equipped with ladders, tools, and material, has proved effective in Cincinnati

Double-width traction wheels enable this heavy tractor to travel over soft ground without sinking in the earth

The lock-bar shown here is the latest invention for foiling the automobile thieves. It is simple, easily attached, and permits the moving of the car in case of fire

To remove the cylinder-head turn the eccentric lever fastener of the claw-clamps and slip off pins

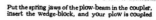

Put the spring jaws of the plow-beam in the coupler, insert the wedge-block, and your plow is coupled

To increase the traction power of small cars pulling a heavy load, they may be coupled by loading the forward end of each car upon the truck ahead

to Automotive Problems

are doing to add to the efficiency tractor, and pleasure cars

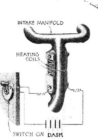

A new electric heater for the intake manifold that will aid in starting the automobile engine

This device indicates automatically the air pressure in your tire

A rolling-mill in Ohio is located about five miles from the nearest city. To solve the transportation problem, an automobile truck and trailer are used to carry the men from the end of the car-line to the mill and back again

This shock-absorber for Fords has three coiled springs and four of them are used for each car

If you use one of these pneumatic jacks it will be possible for your engine to lift the car by air-pressure

You can try the petcocks of your oil channels without any discomfort by rods extending to the running-board

This new type of automobile exhaust-gas whistle requires no electric current to operate it. It is mounted on the exhaust manifold, and is set in action by a wire from the panel-board

This collapsible cot with a mattress and tent roof, is likely to appeal to automobile tourists for economic reasons

To take up the wear of the worm-gear of the steering device, a mechanic has invented this adjustable eccentric shaft bearing

How the Cipher Expert Works

The message that reaches him may be a wild jumble of letters and figures, but it soon gives up its meaning

By Charles A. Collman

THE military cipher expert gains only a small share of war's honors or glory; but in the quiet of his office, his swift, subtle, and scientific efforts in deciphering intercepted dispatches and orders many times bring defeat and disaster to the enemy.

Other branches of the military service, engaged upon more active duty, gather the material, which is laid later before the man whose task is the solution of ciphers abstracted from enemy communications. The scout patrolboat, silently plowing the waters off our Atlantic coast in its quest for submarines, suddenly picks up from the ether the following message, which is recorded by the receiving instrument of the wireless operator on board:

3 S I U O W S E S I L D R Z L A L A
N A B D E n L Y D a N L F C O E T I
U O R G H F K A I R L M T E M T

The wireless operator instantly perceives that this is only too evidently a cipher dispatch of the enemy; and, since there are no other enemy craft suspected of being in these waters, it must be a communication sent by one U-boat to another.

No Cipher Indecipherable

The message is at once rushed by the quickest available means to the nearest office of the Intelligence Department, where it is submitted to the scrutiny of the cipher expert. It is accepted by him as an axiom that no practicable military cipher is mathematically indecipherable. But it requires a remarkably long time, as a rule, to decipher even the simplest kind of message. Therefore speed is essential, so that advantage can be taken of the news contained in this secret communication before it is too late.

So the cipher expert starts to work immediately. His office is amply supplied with all the known data and material by means of which the most up-to-date cipher communication may be solved. He is provided with tables of frequency of the language of the enemy country, covering single letters and digraphs of duals or pairs. He knows that ciphers are divided into two general

classes—the transposition cipher and the substitution cipher; and his first endeavor is to learn in which class this dispatch belongs.

How the Human Mind Works

Substitution ciphers may be made up of substituted letters, numerals, conventional signs, or combinations of all three. The human mind works along the same lines always. The inventive man creates, but his supposed invention is, in principle, many hundreds of years old. For instance, the basic system invented by the Abbot Tritheim (1462-1516) represents the foundation of the modern substitution

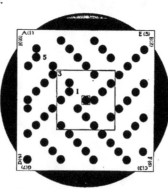

The pattern cipher cannot be decoded without the aid of this ground card

The perforated pattern, which, when placed on the ground card, discloses only the letters to be read

ciphers which are most widely used. Transposition ciphers are limited to the characters of the original text, which are rearranged singly according to some predetermined method or key. This method of secret writing is used to obviate the difficulty of deciphering substitution ciphers, since it is essential that the recipient of the message solve it quickly.

Fortified with his knowledge of the various known cipher systems, the cipher expert closely examines the message that has been submitted to his scrutiny. Much depends upon this initial scrutiny, in which all of his natural powers and talents are called into play. He soon makes up his mind that his frequency tables are of no use in this emergency, since this is not a substitution but a transposition cipher. Furthermore, he recognizes it as a special type of the latter class, from certain plain indications, such as the use of the figure "3" and the junction of the letters "Da," which represent a key. In fact, he suspects that he has before him a form of transposition cipher known as the "net," "pattern," "stencil," or "lattice" cipher.

As part of his office equipment, the expert is provided with most of the known forms of "nets" or "patterns," which consist of perforated squares of tin or cardboard. In this instance he discerns the use of the medium known as Fleissner's improved pattern cipher, in which the pattern or stencil called into play has fifty-seven small round hole cuts out of or punched into the carton, each of which has room for one letter or number.

The Fleissner "Pattern"

The pattern is a perfect square, and the perforations have been made according to certain mathematical calculations, so that, when the pattern is placed upon a sheet of paper, it may be moved in four different positions, and, letters having been written each time in the openings, the entire pattern surface will be covered with a cipher of fifteen lines having fifteen letters each, or 225 letters in all.

The reader need only refer to the accompanying chart to fashion for himself a pattern out of a sheet of common writing paper, when the problem will present itself to him in the most simplified form. The entire pattern surface is divided into squares, running from No. 1 to No. 7, and the size of the communication will depend upon the squares selected, as follows:

The entire pattern surface, that is, to the seventh square, presents 15 x 15 openings, or 225 letters

	Openings	No. of Letters
Sixth square	13 x 13	169
Fifth square	11 x 11	121
Fourth square	9 x 9	81
Third square	7 x 7	49
Second square	5 x 5	25
First square	3 x 3	9

How to Use the Pattern

In the corners of the pattern are indicated the four different positions in which it should be moved; namely, *A*, the first position; *B*, the second; *C*, the third; and *D*, the fourth. It is possible, also, by reversing the surface of the pattern, to obtain four additional positions (*E, F, G, H*); but, since no communication of such extended length is under consideration, this matter may be left to the further experimentation of the reader. The pattern should be moved always from left to right.

From the above sum of the letter openings should be deducted the central perforation, which is not intended for the substance of the cipher message. This opening has a special significance, since here is placed the capital letter (*A, B, C,* or *D*) that indicates the position of the pattern in which the enciphering of the message was begun.

The cipher expert, bearing these facts in mind, recognizes from the use of the number "3" that the third square of the pattern has been adopted

in writing the cipher, and verifies this fact in that exactly forty-nine letter openings have been used, although in two instances two letters each have been placed in the openings.

Shifting the card about according to well known rules, he solves the "pattern" cipher

Now, in beginning his work of deciphering, the expert draws on a sheet of paper a square that corresponds in size with the third square of the pattern, and fills it in with 7 x 7 small circles in positions corresponding with the pattern perforations.

The next step that the expert takes is to write the letters and combinations of letters of the hostile radio message, which results in a table of seven lines, of seven letters each, as follows:

S	I	U	O	W	S	E
S	I	L	D	R	Z	L
A	L	A	N	A	B	D
En	L	Y	Da	N	L	F
C	O	E	T	I	U	O
R	G	H	F	K	A	I
R	L	M	T	E	M	T

Now the decipherer places his pattern upon this table of letters, and discovers that the letters "Da" fit into the central hole. So it is evident to him that the writer of the message began his communication with "D," the fourth position. He therefore arranges the pattern so that "D(4)" appears in the upper left-hand corner, and the openings in the pattern disclose to him the following series of letters: S O R Z A L C T O K L M. This does not make sense, and the expert realizes that the writer adopted position "D" for purposes of mystification. Further observation inclines him to the belief that the use of "Da" as a key indicates that position "D" was utilized merely as a "stall," and that the real message was begun with position "A."

The Message at Last

Turning the pattern into the next position, with "A" in the upper left-hand corner, he reads as follows: "Will Be N N E Fire—" He perceives that he has hit upon the proper solution of the cipher, and, turning the pattern to the left again, to position "B," then to position "C," he obtains the remainder of the message, which reads: "Will be N.N.E. Fire Island Light, Tuesday, four A.M."

All is now known. The proper naval authorities are notified, the machinery of the coast defense is set in motion, and on the day, the hour, and in the position indicated, two U-boats are captured or sunk. And perhaps the cipher expert gains at least part of the credit of the achievement. It may be found interesting to construct a pattern such as the one presented out of tin, cardboard, or even writing paper. The perforations can readily be cut out with a penknife.

Hello, Central! Give Me Berlin

A MODERN army without telephone service is in about the same fix as a big city would be if all the wires should go dead. It follows that every effort must be made to keep the telephone service in the field at top-notch efficiency, and therefore daring linemen and operators follow close at the heels of the troops.

On the fighting line "central" is housed far underground or in carefully camouflaged shelters; but the larger stations back of the front are usually safe from all but aerial attack, and they set up their switchboards in the comparative comfort of a large tent or, better still, an abandoned house.

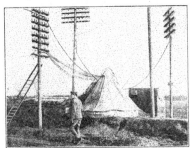

At least, an army "central" isn't asked foolish questions, nor does he have to "listen in" to his gossipy neighbors

In addition to seeing that messages go to the right places, the telephone squad must watch out for breaks and make quick repairs. Formerly the squad had to be on the alert for an enemy "listening in"; but in the latter part of the war the lines were so laid that the use of spy microphones by which messages were stolen from the wire was almost impossible.

The army central's task, while made less hazardous by peace, is none the less strenuous. The title of this article may literally be lived up to, for the army centrals will have a large part in aiding the work of food distribution among the starving enemy.

Now They Motor to the Round-Up

T HE time was, and not so very long ago, when the man who lived on the plains would jump on one of his cow-ponies and cut across the pasture-lands to the round-up. Farmers nowadays go to different round-ups, and travel in an entirely different manner. The pony is left to nibble alfalfa while his master takes his place at the steering wheel of a modern motor-car and skims along good roads to the place of the round-up, where at experiment stations he devotes his energies to learning the best methods for getting hogs and sheep and cattle in prime condition.

Great Guns!

A N entertaining game for children bears the appropriate title "Great Guns!" It consists of a map of Europe in a frame, with a slit in each country to receive a "gun." The object is to coax the guns into their appropriate slots.

Oriental Monsters

W HEN you are in Japan, be careful not to order two crabs for lunch, because two Japanese crabs would keep you eating continuously for at least two days. Rather, ask for a crab's foot, and you will have plenty.

· Japanese crabs, unlike the people, grow to a large and ugly size. If you were to stretch out the two longest feet of one of them and stand him up on one pair of nippers, he would be at least four feet tall. If the power of each pair of nippers corresponds with the length of the leg to which it is attached, a Japanese crab on the end of your toe would most likely mean its loss.

The smiling little Japanese lady has evidently not yet been bitten.

Removing the German Germs

C ANADIANS found this queer instrument outside a one-time German dug-out. When they turned the handle, blasts of impure German air gushed forth, while pure air poured in. It was a German machine for drawing bad air out of dug-outs, worked on the vacuum principle.

Wayside Jewels

B US-RIDING in San Francisco must be a wild, precarious undertaking, for the bus company there has just erected a box-safe in which to place the money and jewels of venturesome joy-riders.

The automobile starter, a very earnest man, is here shown locking up the safe after a bus-load has handed him their valuables. He alone knows the combination of the safe and is responsible for all contained therein, so he needs must be a worthy person. Not only worthy, but brave; for some dark night, while waiting for the return of the riders, he may find himself surrounded by passing burglars. As to whether their jewels are safer here than on the bus must be decided by the owners.

Here's Another Record-Breaker

IN a moment of great patriotic fervor, five hundred tailors, residing mostly in Chicago, decided to build this flag. That it is enormous can readily be seen. But not only is it large—it is the largest ever made. These fervent five hundred carried on a thorough investigation of flags and their sizes before they started work on this one.

They learned that the champion was seventy-five by one hundred and fifty feet, and promptly decided to make theirs eighty by one hundred and sixty feet. And so this enormous creation is nearly thirteen thousand feet square. Each strip is six feet wide and each star is five feet tall. It is so well put together

that there is no "wrong" and no "right" side—both sides are the same. And the colors, too, are "fast," so that no thunder shower will make it "run," and no sunshine cause it to fade.

In spite of its great size, the whole thing weighs only four hundred and fifty pounds,

and is very easy to unfurl. When it was finished, the tailors dedicated it to their fellow workers who dropped their shears to take up guns. It made its debut from the City Hall building in Chicago on Labor Day.

The affair was most impressive. A band of eight hundred sailors played the national anthem as the flag was unfurled. Then two hundred thousand people marched by and saluted it.

It has since figured as the star performer in many patriotic celebrations. Once it was the grand finale to a sham No Man's Land battle which took place at a war exposition at Chicago.

A Cafeteria for Pigs

GORGING continually from morn till night brings unto a pig peace, which in turn causeth fat. Since a pig's function on earth is to grow fat, food should be ever-present in his sight. That this might be, some Ohio farmers built a self-feeder by means of which pigs can eat as much as they liked.

The self-feeder is a small house in which pig feed is stored. Outside of the house there are small compartments in which different kinds of food are always present. This is made possible by adjustable slides which connect the compartments with the food-bins within. A steady flow of food slides down from the bins into the compartments.

Piggy eats a little corn, and then some barley, and then something else. In spite of his greed, he knows what is good for him and never overeats.

Mosquitos Drove Them to It

THESE men are not playing bugaboo with this soldier, nor are they modest Turks veiling their blushing faces, nor yet again are they ghosts that walk by night. They are sane, normal Italian soldiers very sensibly covered with mosquito netting prior to turning in for a night's rest. The comfort of screened windows was not theirs in their rude homes on the Piave, and Austrian mosquitos had been in the habit of feasting greedily on their tired bodies as they tried to sleep. On this particular night, the pests probably slunk away baffled.

Not at Home to the Enemy

IN the earlier days of trench raiding on the western front, a young officer who had been studying medieval methods of defence decided to apply the idea to his trench section. The result, shown in the picture, was generally adopted along the British front.

The V-shaped gate swung outward on hinges at the top. When the Boches were behaving the gate was held up by a chain, permitting free passage into the adjoining trench. But in a raid the guard stood ready to drop the barbed-wire protection and secure it with an iron stake.

An Electro-Magnetic Catapult to Give Airplanes a Good Start

IF an airplane is shot from a catapult instead of starting entirely by its own power, not only can it be started flying more rapidly and in a smaller carriage space, but it can be made to carry nearly ten times more load. And if the catapult is built on the electro-magnetic plan it can be made to serve for airplanes of different design and size, including hydro-air-planes. These are the ideas that prompted Godfrey Lowell Cabot, of Boston, to design and patent a launching mechanism of the kind mentioned.

The catapult takes the form of an electrically driven carriage receiving current from the rails on which the wheels travel, and completing the circuit by means of a third rail between them. A brush attached to one wheel axle takes current to a motor geared to the front wheels, and also, by another wire, to the electro-magnet. The second binding-posts are secured to shoes contacting with the third rail.

The third rail does not begin at the starting end of the track, but a little farther on, the idea being that the carriage and aircraft shall first be set going by the propeller screw, to make sure of the engine being in order. The carriage and aircraft are hurled forward together, and a speed is soon provided at which the aircraft is sure to fly.

An indicator shows the aviator when that moment has arrived, and he signals to the operator of the launching machine, who at once cuts off the current by means of the rheostat-controller. The magnet lets go, and the aircraft sails into the atmosphere.

The carriage is brought to a stop by automatic brakes.

How to Keep a War Map

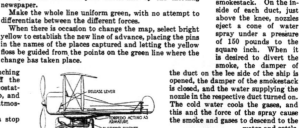

FIRST, get a good map, preferably in black and white, with light blue for the rivers. A recent map prepared by the National Geographic Society can be purchased for one dollar. Then, procure some flat-headed colored pins. Next, take some highly colored floss or worsted—green, for example—and establish your battle-line by the morning newspaper.

Make the whole line uniform green, with no attempt to differentiate between the different forces.

When there is occasion to change the map, select bright yellow to establish the new line of advance, placing the pins in the names of the places captured and letting the yellow floss be guided from the points on the green line where the change has taken place.

That Betraying Column of Smoke— Force It Down to the Water

THE bridge of an ordinary cargo vessel rises to a height of from 33 to 35 feet above sea-level, and towers above the horizon visible to the observer on the bridge of a submarine at a distance of ten miles. As a matter of fact, however, cargo ships of the average type usually betray their presence at a very much greater distance because of the smoke that belches from their smokestacks.

As early as thirty years ago, Sir Alfred Yarrow, the famous English naval architect, originated a method for driving the smoke from steamships downward. The necessity of giving greater protection to cargo ships passing through the danger zone during the war caused a revival of interest in that plan, and extensive experiments were made proving the effectiveness of the method.

From the base of the smokestack, at an elevation of about six feet above the deck, horizontal ducts are carried to the gunwales on each side of the ship. The outer half of these ducts slope down at an angle of about 45 degrees. Normally these ducts are closed by dampers on the inside, close to the smokestack. On the inside of each duct, just above the knee, nozzles eject a cone of water spray under a pressure of 150 pounds to the square inch. When it is desired to divert the smoke, the damper of the duct on the lee side of the ship is opened, the damper of the smokestack is closed, and the water supplying the nozzle in the respective duct turned on. The cold water cools the gases, and this and the force of the spray cause the smoke and gases to descend to the water and settle.

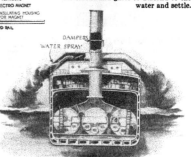

The inventor of this starting apparatus claims that, besides flying more quickly, the airplane will carry a greater load

Smoke from an ordinary cargo vessel can be seen for a distance of seventeen miles. Here it is forced down by water spray

. Ice Gorges in the Mississippi River Grind Helpless Ships:

A Winter's Tale

The man who named this type of boat was surely tempting fate. Fate accepted the challenge, and here **is the result. Horatio C. Wright has** completely lost his **starboard side— lost it to a barge**

This barge was **nestling comfortably in its bunk** at Cairo, Miss., when along came the ice and whisked it away to Norfolk.

These barges, which succeeded in running the gauntlet, emerged from the struggle **sadly dilapidated**

THE Arctic region is not the only place where people can see huge masses of snow and ice heaped up. On our own Mississippi River in the winter-time there is a spectacular display of icy things, with the additional attraction of several wrecked ships and barges. For the severe ice conditions due to the bitter cold play havoc with the river boats, throwing them about in most undignified fashion.

The region around Cairo, which lies at the junction of the Mississippi and Ohio rivers, suffers most. But Vicksburg, Memphis, and St. Louis are by no means overlooked. Early in December the mercury usually slips down below the zero mark, and the river starts to freeze. Then come days of sleet and snow, when all the river boats are frozen in. All this time the water is slowly rising.

Each day the weather grows colder until the thermometer reaches down toward twenty degrees below. Sometimes the river freezes solid from shore to shore. The rising water loosens the ice, and then the ice lets go in huge masses. Barges are dragged away from their moorings, carried down the river, and sunk. Boats are torn to pieces; others are picked up, carried down the river, and sunk.

The ice shows no favoritism. Important government boats receive the same treatment as tug-boats. One winter four large packet-boats, worth half a million dollars, were sunk in five minutes.

This riot usually lasts for several weeks. When, finally, its energy is spent and the ice calms down, the river is full of wreckage. Some halfsunk boats are floated again, and some of the barges are saved; but so many boats are a total loss that this ice spree costs the owners more than a million dollars each year.

There seems at present to be no way of protecting shipping from the fury of the ice.

Putting the Motorcycle on Skates

A MOTOR-WHEEL for skaters, invented by Thomas Avoscan, a citizen of Switzerland, enables lovers of skating and motorcycling to combine the two sports, and to enjoy both at the same time. The invention consists of a tubular frame with an engine-driven wheel at one end, and a sled with a single runner supporting a seat at the other end. The drive-wheel has projections on its tire to give it a firmer grip on the ice. A gasoline or other engine may be used to supply the power.

The rider sits on the seat attached to the runner at the rear end of the frame, and rides on the ice or snow, or he detaches the seat and, standing on the skates, lets the drive-wheel pull or push him over the ice. The handle-bars for steering the wheel are at the rear end of the frame and the turning motion is transmitted to the fork of the wheel-frame by a cable or chain wound around pulleys.

Two pairs of handles, placed back to back, are provided. One pair, with the handles bent downward, is intended to be used when the seat on the runner is occupied; the other pair, with longer arms placed in a horizontal position, is used when the skates are in use.

When the runner is not used, the rod to which it is attached may be folded, together with the runner, so as to rest on the horizontal top of the frame. When the skating seat is used, the operator rests his feet on footrests provided at the lower end of the brace-rod. These rests have downward extensions at their free ends by which the rear part of the frame is supported when the engine is not running.

One of the mechanical features of this invention is apt to confuse persons who are accustomed to riding bicycles. In steering, the wheel turns in a direction opposite to that in which the handle-bars are turned. The inventor claims that this is desirable to obtain the best results in skating. To produce this effect he has provided a crossing of the steering cable between the pulleys.

All Sorts of Johnnies Will
Come Marching Home

It wasn't an exclusive war. Among the Hun-hunters were these Egyptians who have gathered about the regimental scribe for the purpose of sending the good news home

From India — which, despite German propaganda, was loyal — came some splendid scrappers to make things lively on the French front. In the intervals of fighting the Indians were great card-players; but it is said that pinochle is not their favorite game

China represented the far East on the Western front with sturdy labor brigades. These pig-tailless citizens of the republic are watching a dragon fight that has been staged by their leaders

Little Portugal didn't consider her treaty with England "a scrap of paper," but rather as good reason for a scrap with the Hun. Her soldiers made excellent bombers, as Fritz will admit

South Africa sent colored as well as white soldiers. The huskies in the picture, both white and black, belong to a native Red Cross hospital corps

Lo, the fighting Indian! When these original Americans took the war-path against the Kaiser, they proved to be invaluable scouts and most daring trench raiders

While the Armistice Is On

Our Ally from the East shows his idea of a "standardized" ship

"From Berlin to Paris," said the sign on this German made merry-go-round, but its French captors made the sign read the other way

"Merrily he rows along, Rows along, rows along!"

He paddles his home made canoe while Foch lays down his terms

An old motorcycle engine furnishes the power for this trench-made scenic railway built by some Canadians

With characteristic Yankee ingenuity, he has made a comfortable cot out of materials salvaged from the scrap-heap

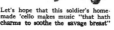

Let's hope that this soldier's home-made 'cello makes music "that hath charms to soothe the savage breast"

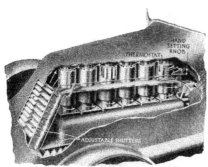

If the thermostat balks, or when there is no thermostat, the driver can regulate the amount of air for cooling the engine

How to Control the Temperature of Air-Cooled Engines

IT is essential for the proper working of an air-cooled automobile engine to maintain a nearly even temperature. Obviously in zero weather less air will have to circulate around the engine to keep it cool than on a hot day.

A recent addition to the large number of devices for controlling the air intake is one patented by Leslie W. Griswold, of Cooper, Ia. The inventor provides for two sets of shutters similiar to the familiar window-shutter. One set is in front of the engine-hood, the other underneath, near the front end.

The shutters are controlled by rod connections with a thermostat, which is mounted on one of the engine cylinders and operated by the heat of the engine. Both sets of shutters can also be opened, closed, or set at any desired angle independently or jointly by hand.

For that purpose a handle with a locking device is provided, which is operated from the driver's seat through an opening in the dashboard. Each set of shutters is controlled by a different motion of the hand-control.

How Yankee Ingenuity Built Roads in Devastated France *

WHEN the American troops arrived in France, they found the military roads near the front in a deplorable condition and entirely inadequate for the safe and rapid transportation of troops and war materials. It became one of first important tasks of the American engineer troops to build new roads and to repair those already in existence.

The rapidity with which the Americans built substantial roads and the thoroughness of their work surprised even the French, who are noted for their road-building. But our engineers had at their command the best labor-saving machinery for road-building. One of these labor-saving devices is shown in the pictures below. It is an apparatus for distributing liquid asphalt for binding purposes over newly finished roads.

The apparatus is mounted in a unit on a special frame which fits over the frame of an automobile truck, and which can be removed as an integral unit in one hour by unscrewing the nuts of sixteen bolts which fasten the frame to the chassis.

The asphalt is carried in a big tank holding 1000 gallons, and is melted and kept in a liquid condition by a system of pipes through which steam, generated in a steam-boiler located at the rear end of the truck, is forced.

Some of the steam generated in the steam-boiler is used to operate a locomotive air-pump at the front of the tank. The air compressed by this pump is fed directly into the tank, and supplies a pressure of one hundred pounds to the square inch. By this pressure the asphalt, which is constantly kept at a temperature of from 150° to 200°, is forced through the nozzles of the sprayer in the rear and distributed over a wide strip of the road-bed.

The air-pump is so arranged that by the changing of a few valves it becomes a vacuum-pump. By exhausting the air from the tank the melted asphalt is sucked into the tank. This is an important feature in view of the great viscosity of the asphaltum and the difficulty of charging the tank with that bituminous substance in any other way.

One of these asphalt-spraying machines greatly expedited the building of military roads in France

You can ferry your car easily across the stream, but what about the car following you?

A Ferry Driven by the Automobile It Carries

THE problem of crossing a deep body of water in an automobile has been solved in a novel manner by L. B. Robbins, of Harwich, Mass. His device makes it possible to run the automobile on a specially constructed pontoon raft, and to drive and steer it across the water by utilizing the power of the car.

The ferry consists of a platform of timbers resting upon two pontoons placed parallel with each other at a distance greater than the width of the average automobile.

The platform is provided with a rectangular opening between the pontoons in which the driving mechanism is placed. Two sets of rollers, mounted on parallel shafts, are placed underneath the platform so that the circumference of the rollers is on a level with the platform. The forward set of rollers has mounted on its shaft a small bevel gear which meshes with a larger bevel gear mounted on the end of a shaft at right angles to the rollers.

This shaft extends rearward in a slanting position under the surface of the water, and has a propeller at its submerged end.

In operating the ferry, the automobile is run on the platform between guide rails, until the rear wheels rest upon the two sets of rollers. The motion of the rear wheels is transmitted by the rollers and the bevel gearing to the propeller, supplying the power for driving the ferry across the water.

The steering device is equally simple and clever. A rod of suitable length is strapped to the front wheels of the car. To the ends of the rod are attached ropes, wires, or cables running over pulleys to the stern end of the ferry, where they are fastened to the tiller.

By turning the front wheels of the car the tiller is moved and the ferry steered as desired.

The inventor has cleverly provided for everything—except the return of the ferry to its starting-point. Perhaps he has faith that there will always be a customer on the other bank, ready to supply the motive power for the trip back.

Removing One of the Winter Worries of Car-Owners

WHEN the temperature outside threatens to drop to the freezing-point, the automobile owner usually begins to worry and to have unpleasant visions of frozen engines and pipes and of large bills for repairs.

An invention that was recently patented by George C. Olmsted, a citizen of Minneapolis, Minn., suggests a possible solution of the problem by providing an automatic device which becomes operative when ever the outside temperature drops below the freezing-point.

The engine is started, and allowed to run a few minutes, until it is warmed up. Then it is automatically stopped by the breaking of an electrical circuit, to be started again after a predetermined interval. Unless the cold is extremely intense, a running of the engine for one or two minutes, at intervals of from five to ten minutes, will result in keeping the water in the radiator and cylinder jackets from freezing.

In addition to the storage battery, an auxiliary battery is also required. The starting and stopping of the engine at predetermined intervals is controlled automatically by a clock-operated disk of insulating material, which has on its surface several concentrically arranged series of spaced contact points.

These points periodically engage a contact arm, thereby opening or closing the circuit.

If your automobile is equipped with this device, even below-zero weather need not cause you any worry.

This simple-looking apparatus makes the unbreathable air of a 20,000-foot altitude possible to inhale. The tube is attached to a bottle containing oxygen

How Aviators Get Oxygen at High Altitudes

THE mechanical difficulties that in the early days of aviation prevented the reaching of heights as great as 15,000 or even 20,000 feet were overcome; but another difficulty had to be solved before such ascensions became practicable. At extreme heights, especially after a rapid ascent, the human lungs do not function properly. They cannot adapt themselves to the sudden change of air pressure, and the aviator is threatened with suffocation.

But this difficulty also was overcome. Each aviator was provided with an extra supply of oxygen upon which he could draw in case of need. The apparatus consists of an Arsonval vacuum bottle enclosed in a metal basket. The bottle is filled with enough liquid oxygen for two persons for one hour at a height of 15,000 feet. When the stop-cock is opened the oxygen passes in gaseous form through a tube connected with the bottle.

The cold produced by the evaporation of the liquid gas is so intense that the gas, if breathed in as it comes from the bottle, would congest the lungs and cause death. To make it breathable it is first conducted through a long pipe coiled around the basket containing the bottle, and then into a rubber bag, from which a tube conveys the gas to the aviator. A second coil, with a rubber bag and service tube, is provided for the use of the passenger.

The Arsonval Vacuum bottle

There is no danger of an explosion should the bottle containing the liquid oxygen be struck by a projectile; but the heat from the burning of the airplane would be disastrous. It would cause the gas to expand and burst the container, and the liberated oxygen would aid in destroying the airplane.

The entire equipment for two persons weighs only about eighteen pounds and occupies but little space in the fuselage of the airplane. In the American army it has recently been ordered that every pilot who goes aloft must carry enough oxygen for from six to eight hours.

How necessary oxygen is to an aviator was demonstrated by the experience of Captain R. W. Schroeder, U.S.A., on his remarkable flight of September 18, 1918, when he broke all altitude records by ascending to a height of 28,900 feet.

Playing Jackstraws with Bales of Cotton

IMAGINE playing the game of jack-straws with bales of cotton weighing 500 pounds each. That game is played daily on a large scale at the New Orleans public cotton warehouse, which has a storage capacity of 200,000 bales. The bales are stacked like bricks in a wall on both sides of wide aisles. Some of the stacks are twenty bales high, and each bale is carefully labeled or tagged, the tag giving the identification number, locality, name of sender, and other information.

Suppose that a certain bale of cotton—which may be at the bottom or somewhere in the middle of a twenty-high stack—is sold. It must be removed without disturbing the bales above, below, and on each side of it.

Formerly it required five or six men and half an hour of time to extract a bale from the stack. Now the same work is done by two men in about three minutes, with the assistance of a special device. This bale-extractor has an A-shaped steel frame with two feet, which are planted against the bales to the right and to the left of the bale to be extracted. Two long arms, curved like those of ice-tongs, are pushed between the bales; the bale to be taken out is grappled by them; and then the bale is extracted from the pile by pulling on a rope or cable which runs over the pulley at the apex of the A-frame and draws hooks and bale toward the top of the A, the feet of which rest against the adjoining bales.

The extractor can be operated by a single man. After the bale is removed, the traveling crane, suspended from the ceiling and operated by an engineer, picks it up and carries it to the loading-place.

The man in the box of the traveling crane controls the power required for extracting the bales

There are four men in the picture, but only one of them is necessary to operate the bale-extractor

Real Flying Dutchmen

Mammoth air-boats that are at home in air and water

By Carl Dienstbach

In action! The flying-boat pro ed no mean adversary for the submarine

THE first man who gave really serious thought to flying across the Atlantic—serious in the sense that he actually built a flying-machine to carry out his intentions—was Glenn H. Curtiss. He decided that his machine must have an enormous radius of action, and to obtain it he considered it necessary not only to increase the size of the airplane, but also to improve its efficiency.

The chief obstacle to an increase of its efficiency was the landing gear. The sheer weight and air resistance of that appendage wasted fuel. But when Curtiss considered the crew, particularly their comfort and safety, he went to the other extreme, and decided to turn the landing gear into a vessel as big as a sea-going launch.

The boat or launch proved to be so heavy that before the machine could get into the air it was found necessary to leave behind most of the fuel. Later he adopted the "sea-sled" type of boat. While Curtiss was still experimenting the world war broke out. He sold his experimental craft—the *America*—to the British Government, which used it very successfully in patrolling the waters around the British Isles.

Curtiss gave to the world a craft that had some of the attributes of both airplane and dirigible, alternately flying and resting on the water. Small flying-boats cannot live on the ocean, and to become relatively seaworthy, seaplanes must have sealed catamaran floats, and the men must be raised high above the waves. Only a mammoth craft, something with a huge hull, something that will transform the flying-boat into a flying galleon, can solve the problem.

These huge flying galleons, as war progress has finally shaped them, rival in beauty the most picturesque old-fashioned ships. With their wings suggesting low-rigged ancient sails, they resemble the pigmy vessels in which the daring pioneer navigators of the fifteenth and sixteenth centuries crossed the Atlantic. Every structure becomes beautiful when once it is perfectly adapted to its purpose. These new flying-boats are beautiful for the very feature that makes them practical—their raised tails. What is it that an oarsman must know before he can safely venture out upon a big body of water? That he must cut the waves. Our galleons of the air are dirigible floating weather-cocks. From their rounded compact hulls waves dash off as harmlessly as from the caravels of Columbus or from Hudson's *Half-Moon*. They head into the wind as quickly as a high-forecastled, high-pooped ship of old, which was likewise a floating weather-vane. Their high tails, when they rest on the water, head them into the teeth of the wind. If the descent is too steep, the nose may be tilted up, yet the high tail drop clear of the water without splashing and without breakage. Alighting is a more delicate operation in a seaplane than in a landplane. It is only too easy to come down nose first. Let the pilot beware lest the prow be caught in the water and the machine turn a somersault.

It certainly seems a most interesting coincidence that ships that in truth navigate alternately the sea and sky have now assumed the identical and fascinating appearance of the legendary *Flying Dutchman*.

The leviathans of air and sea lying in harbor. They suggest the picturesque high-pooped galleons of Columbus's time

Do It with Tools and Machines

Milling attachment for the lathe tool post to cut threads of any pitch

Screw-thread leads may be accurately gaged with the tool post shown below

A universal angle fixture to assist the workman at the milling machine and shaper

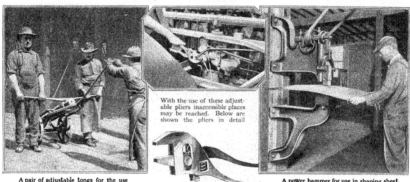

With the use of these adjustable pliers inaccessible places may be reached. Below are shown the pliers in detail

A pair of adjustable tongs for the use of two men in handling heavy car axles

A power hammer for use in shaping sheet metal to be used on automobile bodies

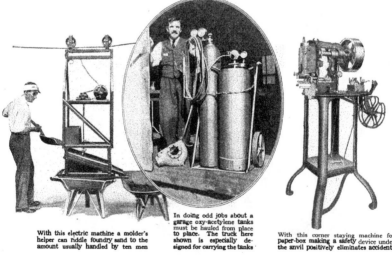

With this electric machine a molder's helper can riddle foundry sand to the amount usually handled by ten men

In doing odd jobs about a garage oxy-acetylene tanks must be hauled from place to place. The truck here shown is especially designed for carrying the tanks

With this corner staying machine for paper-box making a safety device under the anvil positively eliminates accidents

Making Things Easier for the Housekeeper

A tray that can be clamped to the base of a syrup jar, for the purpose of catching the drippings

The adjustable shelf on the baker at the right insures baking food underneath as well as on top

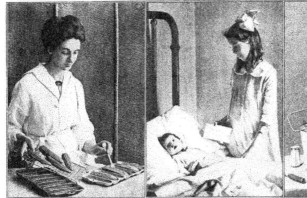

This baking-pan has depressions for baking corn bread in the shape of an ear of corn

A heater to keep the baby warm in bed or in its carriage on a cold day

This strainer can be adjusted without touching the strainer with the hands

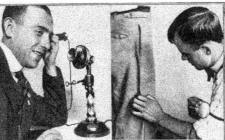

A latch attachment for locking a door with a padlock; adaptable for a barn or shed door

A diminutive hand washboard, which may be used in washing laces, hosiery and other small articles

The hour-glass used for timing the boiling of eggs is here adapted to use as a telephone timer

A bicycle trouser guard and a thick pin-cushion make a convenient place for this tailor's needles

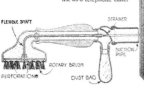

FLEXIBLE SHAFT
STRAINER
SUCTION PIPE
ROTARY BRUSH
PERFORATIONS
DUST BAG

This rotary brush combined with a vacuum suction is used in cleaning seat cushions

A special sewing-machine attachment which darns fine hosiery very satisfactorily

An attachment for keeping scissors in a convenient position for picking up

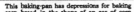

Righting the Capsized "St. Paul"

She is about twice as heavy as any other ship that has been rescued from a similar fate

By Joseph Brinker

WHEN the 13,000-ton American liner *St. Paul* turned turtle at her dock in New York, she settled down fourteen feet into the mud, with her topsides only fifteen feet above the water at low tide. Imagine this huge timber, 535 feet long, 63 feet wide, and approximately 80 feet high, lying on its side in forty feet of water.

If the ship had been a timber of homogeneous mass and strength, it would have been a neat engineering job to turn it over; but the fact that it was a ship, with all its rigging super-structure, hatches, portholes, and heavy weights of engines and boilers concentrated at particular points in the hull, made the successful completion of the task one of the most remarkable in the annals of marine wrecking.

In turning over on her side, the *St. Paul's* masts and stacks and some of her superstructure was crumpled up in a tangled mass like so much paper and sticks. This, of course, had to be removed before any of the actual salvage operations could be begun. Then, there were the naval guns to be removed while the floating operations were under way. To remove the rifles on that portion of the deck out of water was a simple task, but that of getting out those which had their barrels stuck anywhere from ten to twelve feet in the mud was a different matter. It required the greatest skill on the part of experienced divers, since it had to be carried on in the midst of much wreckage.

A Salvage Miracle

With the guns removed, the real salvage work on the ship began. This was done in three steps: first, removing the superstructure and rigging wreckage ; second, righting the vessel to a vertical position ; and, third, floating the ship after righting.

Under ordinary conditions, the removal of the superstructure wreckage might have been a simple job; but in the restricted area of the slip, which was approximately six hundred feet long and two hundred and forty feet wide, it was a most difficult one, particularly as the vessel took up eighty feet of the width of the slip.

In the great mass of tangled stays and rigging, davits, cowl ventilators, and splintered masts, it was most difficult to free the wreckage, especially the crumpled stacks, since each stack weighed many tons and was of such bulk as to be decidedly unwieldy. Skilfully placed sticks of dynamite were employed to blast off the stacks and to free the smaller wreckage.

The next step was to determine how the vessel was to be righted and raised.

Divers were sent into the inky hull of the vessel, and reported that there were several hundred openings in the shell through which water entered and through which the silt and mud from the river bottom was gradually working its way in great layers from six to eight feet thick.

As many as twenty-two divers were used a day, and the work of directing them fell upon Captain I. M. Tooker, superintendent, to whom great credit is due for the water-tightness of the hull during pumping operations, and for the absence of serious accidents.

Boilers Found Intact

Besides closing up all of the ports and openings through which water and silt could enter the hull, it was necessary for the divers to make new openings in bulkheads and the like in order to facilitate pumping out the water later on. In some cases, these apertures were made by means of electrically igniting sticks of dynamite placed at the proper locations by the divers. While this method was certain in results, it was not entirely satisfactory in that many of the adjacent ship members and plates were damaged, and to be repaired.

To offset troubles of this kind, Ralph E. Chapman, the salvage engineer in charge, and J. W. Kirk, the machinist foreman, developed an under-water oxy-acetylene flame device which can be employed while entirely submerged.

Showing how tackles attached to the A-frames pulled, while cables from the pontoons lifted, to turn the *St. Paul* upright

In addition to cutting openings in some ship members to facilitate the flow of water once the pumping operations were begun, it was found necessary to erect barriers in other parts of the ship in order to confine the water intended to be pumped out at any one time. Heavy concrete patches and walls were utilized for this work. Luckily, the boilers and the engines were found intact.

The next step was to raise the vessel to a vertical position without moving her lengthwise, so that her bow or stern did not foul the dock piles fifty feet away. On account of her mass of thirteen thousand tons, it was evident that some new means would have to be resorted to in order to right her successfully. After much study, it was decided to attempt the work by means of block-and-falls. This is fully described in the illustration on the opposite page.

While both the A-frames and the pontoons were being placed, dredging was being carried on at both sides of the hull where it lay in the mud.

Righting the Vessel

The righting operations by means of the pontoons and by pulling on the A-frames were not begun until July 22, although the boat sank on April 25, the long interval being due to the great amount of necessary preparatory work. Once started, however, it took only six days to right the vessel from the original angle of 73 degrees to which she sank, to 27 degrees, where the pontoons were no longer effective. The pontoons were then removed, and two cofferdams placed on the sides of the vessel amidships. These were necessary to prevent the water rushing into the hull, since the boat was still resting on the mud and the high tide was above the decks. The next operation was performed by two large floating derricks, which took hold of the chains on the port side of the hull that had been previously attached to the pontoons, and exerted a lifting and heeling force while the pumps were removing some of the water from within the vessel's hold. This work brought the ship to within sixteen degrees of the vertical, when the A-frames were removed, after which four floating derricks steadied the ship, while the pumps continued their work until she floated free of the mud.

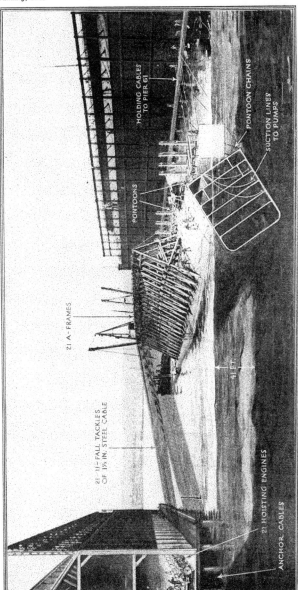

ANCHOR CABLES

21 HOISTING ENGINES

21-11-FALL TACKLES
OF 1½ IN. STEEL CABLE

21 A-FRAMES

41 FT.

PONTOONS

HOLDING CABLES
TO PIER 61

SUCTION LINES
TO PUMPS

PONTOON CHAINS

DREDGE OUT SECTION

DERRICK
WRECKER

A-FRAMES

WIRE CABLES

When the *St. Paul* had been raised to an angle of 27 degrees, large floating derricks took the chains attached to the pontoons, and continued the work of lifting, while pumps removed water from the ship

(STANDING FLOATS)

HT

40 FT. 300 FT.
COFFER DAM

Two cofferdams were put in place to prevent water from rushing in through the open decks at high tide when the ship had been raised to 16 degrees, but was still resting in the mud on the bottom

Methods Used in Righting the "St. Paul"

The great ship was literally pulled upright by twenty-one block-and-fall tackles anchored to ten-ton blocks of concrete, and running to the tops of twenty-one A-frames, each about thirty feet high, made of timber and steel plates, and placed amidships transversely of the hull, as shown above. One upright of each frame was placed at the line of the upper deck, so that the pulling strain would be transmitted to the deck-beams in order that no shell plates might be torn off. The foot of the other leg of the frame was held in place near where the side-plating of the ship joined the bottom by means of lugs attached to the hull.

The simultaneous operation of hoisting engines, to which the tackles were carried, exerted a pull that gradually brought the 13,000-ton mass upright. The righting of the vessel by the pull on the A-frames was aided by a pull from the other side exerted by chain cables attached to the port side of the ship and leading to four large pontoons placed between the *St. Paul* and the pier. The cables extended up through wells in the pontoons, and were carried vertically on wooden uprights. Hydraulic jacks mounted on the pontoons themselves took up a notch at a time during the lifting operations.

Each pontoon was pumped out to give it more lifting power, and was so anchored to the pier that it could not move toward the *St. Paul* as the vessel turned.

Nature Supplied the Material

FOR reasons best known to himself, a wealthy Cincinnati man desired to have built on his estate a log cabin, to be constructed, as far as possible, from the rough materials that nature provides. He got what he wanted—more or less. For about two years the builder assigned to the job wandered around the woods and shores, plucking branches and gathering stones. When he looked over the accumulated material, he decided to build a three-room log cabin.

The outside looks like a regulation log cabin much in need of a good finishing off—all except the chimney, which, right in its center,

has a most puzzling hole. This turns out to be an entrance for the use of frisky ones who prefer that way of entering the house to the front door. They run the risk of being covered with soot, but then, there's no accounting for tastes.

In the entrance-hall stands a de-

cidedly rustic chair. A few trees are scattered around (one of them serving as a staircase support to the regions above); there is an ordinary glass window; and, lastly, a little electric light bulb nestling in the ceiling. Of course the window and the electric light aren't strictly in accord, but even naturalists must have some conveniences.

As for the upper regions, they are distinguished by more windows, the chimney-door, another bulb, and some young trees tacked on the walls.

For the artistic result, little can be said; artistic ideas vary with the individual.

The second floor of a Cincinnati man's "primitive" log cabin, the exterior of which is shown above

This is the entrance-hall, with a tree serving as the stairway. An electric bulb nestles in the ceiling

Knock-Down Snow-Sheds to Protect Miles of Track

KNOCK-DOWN snow-sheds of concrete construction are the latest thing in the railroad world. A bitter fight to keep the Union Pacific tracks clear started with the storm that began December 20, 1916, and was prolonged with increasing intensity through January and February. The finely powdered snow swept downward from the mountains, carried by a gale that often attained a velocity of sixty miles an hour. Whenever a train stopped for coal or water, the snow would pile up around the wheels, blocking the train.

Within five minutes after a snow-plow had fought its way through the drifts, all of the labor would have to be done over again. For six weeks zero was the highest mark.

Driving a rotary plow into one of these frozen drifts was like driving a

motor-truck into a stone wall. Temporary snow-sheds were built, but it was evident that more substantial structures would have to be planned.

As a result, new sheds were decided upon. Briefly, these sheds rest on concrete piles or pedestals. A-frames serve as braces on the sides. Reinforced concrete girders serve as roof supports. Slabs of reinforced concrete fill in the sides and form the roof. The piles were driven from fifteen to thirty feet.

Because of their uniform construction, they can be moved to any location desired.

The A-frames are nine feet wide at the bottom, and twenty feet high. The concrete slabs are 2½ inches thick. Light is provided by omitting the top row of slabs on the leeward side

Minds as Sharp as Bayonets

How the Library War Service provides
for the education of our fighting men

The boxes used for shipping books overseas are provided with a shelf, so that they can be set up like a sectional bookcase

THE complexities and constantly changing character of the war, and the necessity for the rapid training of specialists in every branch of service, created an emergency in the matter of supplying text-books. Many subjects developed so fast that print could not keep up with them—for instance, aerial photography and camouflage. Army liaison, never before employed on so large a scale, had been written of scarcely at all.

The Library War Service of the American Library Association engaged to furnish any published technical work or text-book that might be needed in army camps, war schools and colleges, or naval establishments, in this country or overseas. It could not very well furnish books that have never been written, but it could do the next best thing. It could furnish the technical and scientific data from which such text-books might be compiled. If, for instance, the latest work on camouflage were hopelessly out of date, the Association would get together all of the available books on animal protective coloring and color-photography. From reference books of similar value the army experts prepared their own text-books in mimeographed form.

Arts of Peace Applied to War

Many of these army text-books concerned the adaptation of familiar arts to army uses. A book on motorcycles that was prepared at Camp Joseph E. Johnston, Florida, included a complete manual of instruction for the convenience of the despatch riders, with very complete chapters on the personnel, equipment, and organization of motorcycle companies in the field.

In a whole library on the subject of printing there exists no book especially designed to fit the needs of army printers.

War and the army have a nomenclature all their own, with scores of abbreviations unfamiliar to civil workers. An army print-shop is different from an ordinary composing-room. The printer had to learn the etiquette of war. He had to know, for example, exactly in what type to set a general order, and by whose authority an officer may have his name and telephone number on a letter-head.

When the libraries were first planned, it was thought they would be used chiefly for recreation. It was found that the most regular readers used them for study purposes. This meant a rapid reorganization of the book collections, involving the purchase of 600,000 technical and military books. Gift books to the number of 2,600,000 volumes, the donations of the public—largely in the classes of fiction, poetry, history, biography, and travel—were also placed in service.

Distributor of Literature to the Army and Navy

Enormous quantities of the "Burleson" magazines are used in barracks, hospitals, recreation huts, and on troop trains. The sorting of these magazines entails considerable labor, for men in camps are very like men outside—they want new magazines. Millions of these gift magazines were distributed, and the American Library Association also entered subscriptions to eleven magazines to be supplied to all Y.M.C.A. and K. of C. huts where reading facilities are provided, as well as to the libraries proper. These magazines did much to supplement books in providing up-to-the-minute information on technical topics that were still in a state of flux.

Arrangements were made with both Admiral Sims and General Pershing whereby the American Library Association became the official distributor of literature to Americans overseas, and fifty tons free cargo space was allotted each month for books for our boys.

Thousands of reference works and text-books were furnished to the army and navy by the American Library Association, but many more are needed. There are constant calls for books on mathematics — arithmetics, trigonometries, and geometries. Most of these books had to be purchased, but some were donated.

For the first time in history, the intellectual needs of a great fighting force were scientifically considered and met.

This library interior at Camp Jackson is typical of the buildings which the American Library Association erected in every large camp. In the oval above—Navy boys studying electricity take their library books to the switchboard, in order to combine theory with practice

Once installed, this system of house-heating with kerosene continues to function without requiring further attention. The descriptions on the picture explain the details of the apparatus used in this heating system

You Can't Get Coal?
Heat Your House with Kerosene

THE difficulty of obtaining coal has induced many house-owners and manufacturing concerns to substitute kerosene. The system of house-heating with kerosene is easily installed in any house equipped with a steam or hot-air furnace. Aside from the removal of the grate bars, no change is necessary in the furnace. Any good mechanic can install the system in a few hours.

The apparatus consists of a kerosene-tank connected by a narrow feed-pipe with the combustion-chamber, a motor-driven blower which atomizes the oil and blows the spray into the combustion-chamber, a pilot light to ignite the oil spray, and a thermostat for regulating the fire according to the desired temperature. When the temperature of the room reaches the point at which the thermostat is set to operate, the electric circuit to the motor is broken, so that the blower stops feeding fuel into the combustion-chamber and the fire goes out.

When the room temperature drops and the thermostat again makes the electric contact to the motor, the fuel is ignited by the small pilot light in the botton of the combustion-chamber.

This method of using kerosene

instead of coal as a fuel in house-heating plants undoubtedly offers many advantages. It is much cleaner than coal, and, once it is installed and set in operation, requires practically no attention. There is no smoke, soot, or waste, no ashes to be removed, and the temperature is regulated automatically. But the system is not economical. The cost is as great, probably greater, than the cost of coal. Another drawback lies in the fact that electric current is required to drive the motor and illuminating gas for the pilot light.

Making the Radiator
Easy to Look At

FOR the benefit of fastidious persons who object to radiators because they disturb the artistic harmony of beautifully furnished rooms, a Western manufacturer has placed on the market an imitation woodwork cabinet for disguising the naked ugliness of steam radiators. It is a sort of artistic camouflage that in no way interferes with the usefulness of the heaters.

These cabinets, which are made to order on any style, are of heat-insulated metal, painted to imitate the color and grain of any kind of wood desired. The cabinet is open at the bottom and is placed over the radiator. The cold air from the room enters at the bottom, is heated by the hot steam coils, and passes out into the room through a series of grilled openings near the top of the cabinet.

The slits are provided with shutters, which are connected by a bar with the operating arm of a thermostat installed in a specially provided chamber at one end of the cabinet.

The thermostat may be set for any temperature desired, and when that limit is reached it will automatically close the shutters of the slits through which the hot air passes into the room. The thermostat is operated by the expansion and contraction of an air-tight copper or brass cylinder with corrugated, bellows-like sides.

The bellows movement of the cylinder is transmitted by an ingenious system of levers to the rod, which controls the shutters and causes it to open or shut the slits automatically.

You may artistically camouflage your radiator by covering it with a cabinet. The thermostat inside the cabinet will automatically regulate the temperature of the room

Save the Nation's Energy—Fuel

Learn how to handle your furnace and save coal and doctors' bills

By A. M. Jungmann

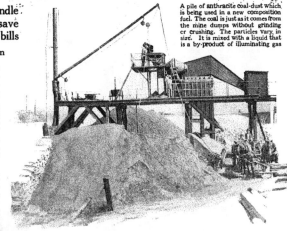

A pile of anthracite coal-dust which is being used in a new composition fuel. The coal is just as it comes from the mine dumps without grinding cr crushing. The particles vary in size. It is mixed with a liquid that is a by-product of illuminating gas

IF you want to enjoy better general health and freedom from colds this winter—save coal. A quart of water evaporated in every room of your house every day will mean a saving of one third of your coal bill. What we really need in our houses is humidity rather than a high degree of heat. Moist air feels warmer than dry air of the same temperature, and moist air has the further advantage of retaining heat.

Professor Ellsworth Huntington, of Yale University, recently made a study of some nine million deaths in all parts of the United States and in France, Italy, and Japan. He also made a survey of fifty million deaths in Belgium, Great Britain, Germany, Russia, Rumania, Spain, and other countries, bringing his investigations of deaths up to the staggering number of sixty million. His study has resulted in the conclusion that an average temperature of 64° F. is the best for the maintenance of health. Professor Huntington also found that a uniform temperature is not so healthful as a variable temperature. He concludes that a frequent fall of temperature, followed by a more gradual rise, is an excellent means of preserving health.

Do Not Heat Sleeping-Rooms

Persons who desire mental and physical vigor should sleep during the winter in rooms in which the temperature ranges from freezing to 40° or 50°. It is not necessary for sleeping-rooms to be warmer than 50° at any time of the day. Windows always should be partly open at night and plenty of fresh air admitted to bedrooms.

The reason people keep their houses at such a high temperature in winter is that the outside air, after it is brought into the house and warmed, usually becomes as dry as the desert of Sahara. This dry air feels colder at 70°

The composition fuel in use. The ash on the floor in front of the boiler shows freedom from clinker and a free-burned granular condition

than moist air would at 60°. By arranging to evaporate water in your house you can save coal because it will not be necessary to keep the air at such high temperatures.

If you are using a steam or hot-water heating apparatus, keep a pan of water on every radiator. These pans should be provided with wicks. There are on the market valves to be attached to steam radiators, which play moisture into the air of the room. If you use a stove, keep a pan of water on top; and if you heat your house by a hot-air furnace, keep a pan of water in the drum, so that the moisture will pass up with the heated air.

Experiments in Ventilation

Professor C. A. E. Winslow, also of Yale University, conducted some very interesting experiments in ventilation for the New York State Ventilation Commission. He found that the latest methods of ventilating rooms by taking air into the cellar, warming it to a temperature of about 67°, and blowing it into the rooms by fans was not so healthful as the ordinary procedure of

The U. S. S. *Gem*, which has been used to test different types of fuel; she has made many successful trips operating on pulverized coal

keeping the rooms at the same average temperature but ventilating them by letting fresh air in at the windows. In the modern system of ventilation the temperature did not vary and there were no drafts. Yet people who worked in rooms under these conditions suffered from colds to a greater extent than those who worked in the rooms where the temperature varied.

Although December and January of last year were unusually cold, the death-rate was low. This probably was caused by the fact that people did not keep their houses heated as much as they have in other years. Shortage of coal, instead of being a hardship, was a blessing in disguise.

Experts in the Bureau of Mines have made some suggestions on as to how to burn coal economically at home. First of all, they lay stress upon the fact that the person who is tending the furnace must acquaint himself very thoroughly with the apparatus. If you want to burn the coal faster, supply more air through the grate. More air can be supplied by opening the damper in the pipe leading to the chimney, by opening the damper in the ash-pit door, or by shaking the ashes down from the grate. If you want the coal to burn more slowly, reduce the draft by closing the chimney damper or the damper in the ash-pit door.

If you will put weather-strips around your windows, the saving in coal will amount to many times their cost

Be sure the glass in your windows is tight. A little trouble in puttying the window-panes will save coal

Learn to Manage Your Furnace

To increase the supply of air over the fuel-bed, open the damper to the chimney or the damper in the firing door or both. Air introduced over the fuel-bed helps in burning the gases and the visible smoke rising from the burning fuel. In order to be really economical, you must study your furnace and find out just how much air it requires; for too much air is as wasteful as too little.

A heavy fuel-bed generally gives more satisfactory results than a light fuel-bed, because, during the long periods between firing, a light fuel-bed is apt to burn down too much. When burning bituminous coal it is better to place the fresh coal somewhat in a heap to one side of the grate and leave a small part of the burning fuel uncovered. If the coal contains a large percentage of slack, this method of firing is particularly satisfactory, because the coal is heated gradually and the volatile matter distils slowly and is given a chance to burn. Before the next firing the coked heap of fuel should be broken and spread over the grate. A new charge is then placed on the opposite side of the furnace.

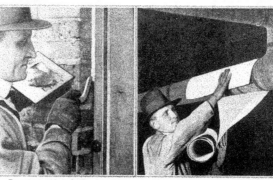

To prevent cold air leaking into the house, calk the doors with an elastic substance that vibration will not effect

If your total coal consumption is fifteen tons a year, you can save at least three tons by insulating the pipes

Mercer P. Moseley, Chief of Conservation, has compiled a list of "don'ts" for coal-users. If you observe these rules you will be able to effect an appreciable saving in coal:

Don't fail to clean furnace before starting fire.

Don't build a fire until necessary.

Don't build a fire larger than is necessary.

Don't fail to make check-draft damper in smoke-pipe do its work.

Don't neglect keeping fresh water in your steam-heater boiler.

Don't fail to keep your kitchen stove clean.

Don't keep your home at over 68° F.

Don't leave your draft open at night.

Don't try to heat all of outdoors.

Don't keep your fire going on pleasant days.

Don't sit in north room when the sun heats the south side.

Don't think it's fur-coat weather when the thermometer is 45° to 50°.

Don't waste water—it takes coal to heat it.

Don't forget that one gas-jet will raise the temperature of a room five degrees.

Don't fail to put up storm doors and windows.

Don't fail to sift ashes.

Don't burn coal when wood is available.

Don't fail to wrap your pipes with asbestos.

Don't fail to keep rooms moist—they heat easier.

Don't forget that moist air retains heat.

Don't forget that dry air causes colds and catarrh.

Don't waste gas—it is made from coal.

How to Bank the Furnace

Do not let your fire burn too low before banking for the night, because it may go out after banking. When banking the fire, leave a part of the surface of the fuel-bed uncovered in order to prevent explosion of the gases rising from the banked fuel. The regulation of dampers after banking

Examine the nails around your window-casing: if any have become loose, you will save coal by tightening them

Take the time to clean your flues out thoroughly at least once a week; soot is a poorer conductor of heat than asbestos

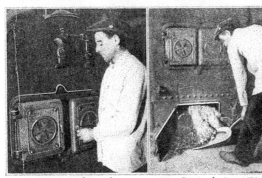

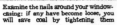

If the air enters from below, in burning soft coal, the oxygen is burned out; the remedy is to let in air from above

Never allow the ashes to get within six inches of the grate; a full ash-pit cuts down the heating efficiency twenty per cent

the fire depends on the amount of air that leaks into the ash-pit, and is something you must determine for yourself. It is generally wise to close the damper in the ash-pit door and leave the damper leading to the chimney partly open in order to prevent coal-gas getting into the house.

Saving Coal in Industry

The best way to start to save coal in boiler plants is to give up guess-work and apply scientific principles. The men who actually burn the coal should be trained in the principles of combustion. No matter how good a man's intentions may be, he cannot save coal for you unless he knows how and knows why it is necessary for him to do certain things in order to accomplish the desired results. Although a very clever fireman may be able to

tell you all about your fire by simply looking into the furnace, you cannot expect every man to do that. The only way you can really know the condition of your fire is by using recording or indicating instruments. Accurate instruments, such as draft-gages, flow-meters, and pyrometers, will help you to handle your coal intelligently and economically.

Some of the things to be guarded against in using coal are clouds of black smoke coming from the stack, which is always a sure indication of waste; the loss due to high percentage of carbon in the ash; and the wasting of exhaust steam. Your steam is a direct product of your coal. There is usually a great deal of economy in using mechanical stokers. Needless to say, it is necessary to see that your entire apparatus is in good condition, that the boilers are clean, that the

baffles are in good order, that the boiler-tubes are not full of soot, and that the flues are air-tight, and everything ship-shape. A leaky flue will cause the burning of more coal than is necessary. Wherever possible avoid turns in your flues. The ideal flues are short and straight. Flues that have turns offer resistance to escaping gases and do not afford the free draft necessary to satisfactory combustion of coal.

Pulverized Coal

Pulverized coal is solving many fuel problems. The ideal condition in which to burn coal is pulverized. With pulverized coal there can be no clinkers and no smoke, and the combustion is complete.

Pulverized coal is ground so fine that a pinch of it between the fingers does not feel gritty. A cubic inch of coal is ground into such fine particles that 95 per cent. of it will pass through a sieve having 10,000 openings to the square inch. The advantage of pulverized coal will become apparent when you remember that a cubic inch of coal has a surface or superficial area of six square inches; but when it has been converted into these small particles its superficial area is increased some 700 times. This means that each one of the two million particles is surrounded by air and, when burned, permits perfect instantaneous combustion. There is no waste. Pulverized coal burns like gas.

Not an Untried Fuel

Pulverized coal is not an untried fuel. Nearly ten million tons of it are being used in the United States each year. In the manufacture of cement six million tons are used; in the production of copper one and a half million tons; in the iron and steel industry, two million tons; in the generation of power, some two hundred thousand tons. The cost of pulverizing coal is not at all prohibitive. It can be prepared at a cost of from sixty cents down to twenty cents a ton. The coal is burned by projecting it into the furnace by means of an air-blast, forming a cloud of thoroughly mixed air and coal.

Where scarcity of labor is a factor to be considered, pulverized coal is a very desirable form of fuel, because one man can handle a number of machines; for the process of burning this form of fuel is almost entirely mechanical.

Pulverized coal is not only suited to stationary plants, but is used with gratifying results on locomotives and steamships. Its use on locomotives is especially promising because of the saving of both time and labor. Burning pulverized coal on a locomotive does away with the necessity of a fireman. The coal is supplied to the fire automatically at the discretion of the engineer. There is no smoke, and no cinders are spilled along the road-bed. When it is considered that one railroad in the United States spent in 1916 $375,000 in settling claims for fires started by sparks or burning ashes along its right of way, you can see what a great advantage pulverized coal has over the usual locomotive fuel.

Used on Locomotives

Locomotives operated on pulverized coal do not have to waste time at ash-pits or in cleaning fires and shaking grates. When the fire is extinguished there is no waste of combustible in the ash, for there is no ash. Inspection can be made very easily. A series

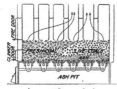

A square fire-pot fired by the coking method

of tests show the advantages of pulverized coal for operating locomotives over ordinary coal.

A Mikado type of locomotive in a fast freight service running over a 115-mile division consumed about 29,500 pounds of coal for six trips, as compared with another locomotive of the same type, equipped for burning pulverized coal, and which consumed 22,500 pounds of coal, or a saving of 23.7 per cent in fuel. A locomotive burning pulverized coal has been in constant daily operation near Fullerton, Pa., since January, 1918, and up to the present has not been laid up for repairs.

The Coal Situation This Year

Experiments in burning pulverized coal were conducted within the last few months on the U. S. S. *Gem*, and proved that, where a steamship had storage-room for that fuel, pulverized coal is eminently satisfactory for use in operating ships.

The coal situation this year is particularly interesting. The coal year begins on April 1. The anthracite

This is the apparatus that is used for burning pulverized coal on locomotives

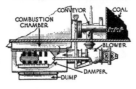

coal consumed for domestic purposes in 1916 was 49,258,000 tons. To this the Fuel Administration has added 2,000,000 tons, bringing the total to 51,258,000 tons, which it has undertaken to supply this year. The first half of the coal year was up on October 1, and the supply for that period was exceeded by 759,136 tons.

Bituminous coal is produced greatly in excess of anthracite; about twelve times as much bituminous as anthracite coal is mined each year in this country.

This year the United States expects to produce some 600,000,000 tons. In the first six months of this year's coal year 37,000,000 more tons were mined than were produced last year in the same period.

Ever since the war began we have been using an increased amount of bituminous coal—an advance of some 50,000,000 tons for each year.

When you consider the

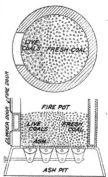

A round fire-pot fired by the coking method

shortage of labor in the mines, the production this year is nothing short of remarkable. If we all do our bit, from turning off the gas or electric light when we leave a room at night, to inspecting the flues in our industrial boiler plants to see that there is no leak, we shall have enough coal this year to keep everybody warm enough for health's sake and to keep the wheels of our reconstruction industries not only turning but humming.

A six-wheel saddle-tank locomotive equipped with a pulverized coal apparatus. It has been in daily operation in Pennsylvania since January 2, 1918. Coal capacity, one and a half tons

Making Periscopes Invisible

IT was the visibility of the periscope, more than anything else, that made an effective war against German submarines possible. A periscope is useless if it is submerged.

The part of the periscope tube that extends above the surface of the water is not more than about two inches in diameter—not much, but enough for the vigilant watchers on the numerous patrol-boats that cruised in the danger zone.

Various methods were tried to diminish the visibility of the periscope. At first they were painted in shades of blue or gray or green so as to make them blend with the color of the sea. Coating the tube with aluminum paint was also tried. But all such attempts at camouflaging were found to have but little protective value. Covering the tube with a green and violet checkerboard proved effective in some cases.

A suggestion that is theoretically good provides that the tube of the periscope be surrounded by a series of conical mirrors, which, when seen from a distance, blend with the background because they reflect the sea.

A washing plant like this, operated by belts from the factory shafts, may save a great deal of money for the manufacturer

A new idea in protecting periscopes

Don't Waste Waste

WE have all been taught to avoid the waste of money, words, and advice; but who ever heard of wasting waste? Yet, waste—that is, the kind that is used to wipe clean machinery, printer's type, etc.—may actually be wasted.

A Western manufacturing concern installed in its factory an experimental plant for reclaiming soiled waste and rags, and found that they could be washed from six to seventy-five times at an average cost of ½ cent a pound. The reclaiming equipment, which consists of a washer, an extractor, and a drying drum, is now manufactured and marketed by that same concern.

These Life-Boats May Be Launched from Either Side of the Ship

WHEN a sinking ship acquires a list to port or to starboard, and the deck is correspondingly inclined, the long line of life-boats on the high side of the deck, all swung from the regulation type of davits, cannot be launched.

The well known fact that sinking ships invariably get a list before the final plunge has been disregarded for years, rendering half the ship's life-boats useless. This terrible condition has recently inspired two Scotch ship-builders to devise means for launching any and all of the ship's life-boats from the lower side of sinking and listed ships.

The boats are stored in a closely assembled group on a boat-deck, preferably amidships adjacent the funnels. Above this concentrated group of boats is erected an elevated steel structure of athwartship extending I-beams, supported on steel columns, each equipped with a chain-hoist trolley for lifting the life-boats from the chocks.

Hauling cables for moving the chain-hoisted boats to port or to starboard are attached to each trolley, and lead outward in opposite directions therefrom to wheel pulleys located at the end portions of the I-beams, and thence inward and down to a hand-operated winding-drum. The ends of the athwartship extending I-beams are also equipped with boat falls and tackle like those on ordinary boat davits. The operation of "getting the boats over" is, first, to raise each boat in turn from its supporting chocks by the chain-hoists. Then the winding-drums are operated to haul the suspended boat outward by the trolley cables to the lower side of the leaning deck, and to the corresponding ends of the overhead I-beams. The final operation of "lowering away" the boat is performed by transferring the support of the boat from the chain-hoists to the falls and tackle.

No matter how badly the sinking ship may list, it is possible to launch life-boats that are provided with this handling apparatus

Exit the Zeppelin Airship: Enter the Zeppelin Flier

Out of failure of the dirigibles grew the largest of airplanes

By Carl Dienstbach

BOMBING planes are primarily weight-carriers. They must be able to fly for hours, which means that they must be able to carry much fuel; they must be able to do much damage, which means that they must be able to transport heavy bombs. It is fortunate that the weight of the crew does not increase proportionately with the size of the machine. The four men of a Handley-Page, for instance, could navigate a much larger craft.

It is easier to state the principle than to carry it out. There is no more difficult engineering task than the construction of a huge biplane. The British deserve the credit of having developed the first big bombing plane —the Handley-Page. To be sure, the Russian Sikorsky—a machine even larger than the Handley-Page—was built before the war, but it failed to meet the test of war because the structural problem had not been solved. Curtiss, in the United States, had also made experiments with large machines. But the fact remains that the British evolved the first practical giant bomber.

Bombing Zeppelins Abandoned

When we consider the difficulty of building a huge bombing biplane, and the failures that greeted pre-war attempts with the Sikorsky and other big machines, it is no wonder that, when Germany determined to intimidate England by bombing London and other British towns from the air, she rejoiced in her Zeppelins. Their radius of action was well-nigh boundless; they could elude early air defense artillery with ease.

There was a rude awakening when, later in the war, one great dirigible after another became literally a flaming altar on which a score of lives were sacrificed. German officers at last saw the truth. The Zeppelin was a good naval scout, but a vulnerable bomber. When it became apparent that the Zeppelin airship must be abandoned for bombing raids over England, the Germans copied the Handley-Page and produced their first Gotha.

But Gothas did not take the place of Zeppelins; and the Germans gave orders for the construction of airplanes of startling dimensions. Aided by the Albatros works, the Zeppelin Company produced a giant bird which marks a distinct advance in the development of the long-range bombing airplane. The new craft's load of bombs is two tons, as against the Handley-Page's half a ton. The span is 135 feet, as against the Handley-Page's 98; the length 94 feet, as against 65; horsepower 1200, as against 740. The speed remains what it was in the Zeppelin dirigible—75 miles an hour—against the Handley-Page's 85.

It is the general practice in large bombing planes to balance the elevator and the ailerons, by which latter side-to-side balance is preserved. In other words, there is usually a balancing surface in front of the pivot-shaft, the object being to make the operation of the control surfaces as easy as possible. In the Zeppelin bomber there is no such attempt at balancing. The conclusion is inevitable that there must have been auxiliary engines (probably electric motors) to operate ailerons and elevator, although French engineers who reconstructed a wrecked machine are silent on that point.

Inside of the wings is an intricate filigree-like wooden lattice-work which is designed to secure great strength with little weight or air resistance, and which reminds one of the bewildering framing of a Zeppelin dirigible's envelope. The spars are of hollow girders. All the other framing is of steel tubing. Wherever the tubes are exposed, so that they must be driven through the air (struts, etc.), they are encased in thin wooden shells admirably stream-lined. The tail-framing, however, is of aluminum, for the reason that the tail, being farthest from the center of gravity, must be extraordinarily responsive.

How the Zeppelin Was Copied

It will be remembered that in the latest Zeppelin dirigibles the Maybach engines were enclosed in egg-shaped cars so as to reduce head-on resistance. In the biplane we find similar "eggs" in which the engines (Maybachs, of course) are housed, while the crew sits in a separate central car. But in the biplane each "egg" contains two 300-horsepower engines, one behind the other. Examine the illustration on the opposite page and you will see that there are both pusher and tractor screws in tandem. This is indeed a bold departure from long accepted doctrines. Why did the Germans violate the rule? Because it was a real advantage to have four separate engines and four seperate propellers and only two engine-rooms. The air resistance of the engines is obviously reduced by half, and the weight of the engines can be brought nearer the center of the machine.

Landing Twelve Tons at High Speed

The greatest problem that confronts the designer of mammoth bombing biplanes is the provision of an adequate landing gear. Here, for instance, is a machine that weighs twelve tons, and that lands at a speed scarcely less than forty miles an hour. Imagine what that means. The Zeppelin biplane's most meritorious feature is its landing gear. Look at the picture opposite. Note how low, sturdy, light, and devoid of drag is its extraordinarily simple framing. Note the wheels arranged in groups of four. This mass of pneumatic tires efficiently supplement the shock-absorbers.

In a small machine a tail-plane such as that on the Zeppelin flier might lead to stalling. But a mammoth machine is free from any such danger because of its momentum and its multiple engines. Without the peculiar tail the main landing gear would have to be very much higher, air-resisting, and ponderous. Because the landing gear is low it becomes necessary to use four propellers of small diameter in place of the two large propellers of the Handley-Page. These make it possible to lower the landing gear.

Clearly, landing on such a gear offers a new problem. Unless the pilot comes down on nearly an even keel, the machine's nose strikes the ground at such an angle that the strain would be more than the front wheels could withstand. This difficulty has been met by the ingenious arrangement of the tail, in which the French engineers who reconstructed the machine beforementioned failed to see anything but a puzzle.

A Novel Anti-Freeze Pipe

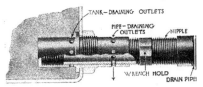

SOMETIMES it becomes necessary to stop the flow of water or other liquid from a tank and to drain the pipe to prevent freezing; or it may even be necessary to empty a tank. In the novel arrangement shown above, the portion of pipe within the tank is closed at the end, but has holes that allow the liquid to flow through the piping when the nipple is not screwed in too far to the left.

In the position shown in the illustration the tank outlets are closed by the nipple, which projects beyond them, while at the same time another set of holes are in the open position for draining the pipe. A slight movement of the nipple to the right or left will close all the drainage outlets, permitting the piping of the liquid to points desired.

It can readily be seen that if the nipple is moved sufficiently far to the right a point will be reached where not only the piping but the tank also will be drained. Thus this simple form of construction permits any one of three separate and distinct operations. The liquid may be piped from the tank, or the supply from the tank may be shut off and the pipe drained, or both tank and pipe may be emptied—all done by a simple twist of the wrist or the application of a wrench.

How to Get Hot Water from Your Camp-Stove

WHEN Francis L. Alsobrook, of Alamo, Tenn., wants hot water, he uses his camp-stove.

The stove is of conical shape, tapering toward the top, ending in a detachable stovepipe for carrying off the smoke in the usual manner. A ring-shaped vessel of galvanized iron fits around the upper part of the stove.

The water-heater (provided with a faucet) has a cover which fits around the stovepipe and has a spring friction catch by which it may be held in any position to which it is raised on the pipe. The water is filled into the heater from the top, after the cover has been lifted out of the way. When the heating vessel is full of water, the lid is shut tight.

The stove may be provided with legs and a bottom, or it may be without legs. In that case, it is set on the ground and the fire is built underneath. The stove has a door for putting in the fuel, preferably wood, and another door underneath, for the removal of ashes.

One of the best features of this stove is that it may be taken apart for moving; for the stovepipe comes out and the heating vessel comes off.

A vessel over the stove heats hot water

The Stars for January

How to find your way in the skies

By Ernest A. Hodgson, of the Dominion Astronomical Observatory

OVER on the northeastern horizon is the shape of the Big Dipper, or Ursa Major—the Big Bear. It seems to be standing almost upright upon its handle. Follow the line of the two outermost stars of the bowl and proceed north for about four times the distance between them. A bright star is reached the—Pole Star, or Polaris. On the map the line to Polaris appears curved, because the stars, which seem to the inner surface of a great globe, are slightly distorted in position in a flat projection such as our map.

Still facing north, look carefully down toward the horizon, from this pole star. The dim outlines of a small Dipper appear, the handle bending the bowl of the Big Dipper. This is Ursa Minor—the Little Bear.

On the opposite side of the pole star from the Big Dipper is Cassiopeia, the Lady in the Chair. This constellation is in the form of a great W; or in its present position, viewed toward the north, it is more properly an M. It is made up of five bright stars—so bright, and the figure they form so clear, that it is probaly one of the best known constellations after the Big Dipper and Orion.

Note carefully the position of the bowl of the small and large dippers. Follow a line from the former to the latter, and continue about twice that distance beyond the Big Dipper's bowl. Here is a bright star, Regulus, at the end of the handle of the "Sickle." The Sickle—or Leo, as it is called—is lying with the crook down and the point up, as we look north and east. In January Regulus is attended by the planet Saturn.

Look over into the northwestern sky. Note the great square of four stars, the two sides of the square forming lines which are directed toward the pole star, as were the pointers of the Big Dipper. This is called the Great Square of Pegasus, although the star in the corner nearest the pole and nearest the zenith belongs to the constellation of Andromeda.

Up near the zenith, above the pole, gleams the bright star Capella, and the attendant stars of the constellation Auriga, the Charioteer. How did the ancients manage to see the form of a charioteer in this group of stars? The shepherds of old had all night to spend looking at the stars, and they had little else to think or talk about. Is it any wonder that they saw strange shapes and had stranger stories about their visions?

Turn now toward the south: There is a sight never equaled in the summer skies. Half way up the southern sky, immediately before us is the constellation Orion. The

STAR MAP FOR JANUARY

How to Find Your Way Among the Stars

This map represents the appearance of the sky at the time listed below. It is specifically arranged for the latitude of New York, 40° N., but is practicable for ten degrees on either side of this latitude. Use the map at the times indicated. Hold it over your head with the north horizon to the north.

December	9, midnight	January 18, 9	P.M.
"	15, 11:30 P.M.	" 25, 8:30	"
"	22, 11 "	February 2, 8	"
"	29, 10:30 "	" 9, 7:30	"
January	4, 10 "	" 17, 7	"
"	11, 9:30 "	" 25, 6:30	"

All times expressed in Eastern Standard Time.

shepherds fancied here a great giant. Betelgeuse is in his raised right arm. Rigel defines his left foot. A triplet of bright stars forms his belt, and, dangling from it, are a line of three fainter stars, the dagger. The lowest of these appears hazy. It is the great nebula of Orion.

Orion is about to deal a blow with his club at the head of Taurus, the Bull. The Bull's red eye—Adebaran—gleams wickedly and its great horns are tossed upwards toward the twins—Castor, and Pollux. For thousands, perhaps millions of years, Orion has kept them safe from their savage foe. In January, Jupiter will add his bright disk to the already bright group of stars above Orien. (See the planet map on page 79.)

The little group, the Pleiades, is known to most people even if they know little else about the stars. The Pleiades, together with Orion and Aldebaran, are mentioned in Job IX, 9, and Job XXXVIII, and the two former in Amos V, 8, which serves to show how well known they were to the people of those days. The Pleiades are referred to as the "seven stars." It was considered a good test of eyesight to be able to count the whole seven stars and be certain of them. With a telescope the group becomes a great mass of stars, the brighter ones surrounded with a great nebulous glow.

Above the Pleiades is Algol, really a pair of stars, one a dark star — or, at least, darker than the other. Once in two days and nearly twenty - one hours, the darker star revolves around in front of the bright one. This causes the combination, ordinarily of the second magnitude, to lower to fourth magnitude for about half an hour. It was called by the ancients "the demon of the slowly winking eye." Mira, in Cetus, is another such variable star.

Other bright stars in the southern sky are Procyon in Canis Minor and Sirius in Canis Major. This last is the brightest star in the sky, and it is the nearest star visible in these latitudes.

In the east, just below the Sickle, is Hydra, the Dragon. Its tail falls far below the horizon, but its head is easily seen to the east of Procyon. By another month it will have crawled forward in the sky until almost all of its shiny length is visible.

Many people do not even know that the stars rise and set as do the sun and moon, and for the same reason. As the earth revolves on its axis, the sun and stars seem to move.

After one becomes familiar with a star group in the southern sky, it is interesting to get two trees or similar marks in line with the group, and after something like half an hour to sight along the marks again to see how far the stars have moved westward—or, rather, how much the earth has turned us eastward beneath them.

Where to Look for the Planets

THE planet map is constructed to show where the planets are for the month of January, 1919. The Sun—the center of our system—has seven other planets besides our Earth revolving about him. If we could take a journey from the Sun outward, with a speed of 186,000 miles a second,—the velocity of light,—we would reach the first planet, Mercury, in a little over three minutes after leaving the Sun. Continuing our journey outward, Venus would be reached about six minutes after the start. Then we would reach the Earth at the end of a little more than eight minutes; Mars at the end of nearly thirteen minutes; Jupiter would require over forty-three minutes; Saturn nearly one hour and twenty minutes; Uranus almost two hours and forty minutes; and finally we would reach the outermost planet, Neptune, after a journey of about four hours and ten minutes.

Now, if we chose our direction so that we headed directly for the nearest fixed star when we first started, and if we kept on at our terrific speed, we would have to journey for about four and one third years to reach this star, Alpha Centauri. It happens that this star is too far south to be seen from these latitudes. If we had chosen the next nearest star it would have been Sirius; but our journey would have been one of 8.7 years before we reached our destination. In other words, if Sirius were to be blotted out today, we would continue to see it for 8.7 years.

To put it another way, if we represented the orbits of all the planets on a page of the POPULAR SCIENCE MONTHLY, we would need to make that of Neptune about six inches in diameter. On the same scale, to plot in the position of Alpha Centauri, we would put a dot out a distance of only a little less than a mile. So our solar system is really a very small family circle, after all.

All these planets shine only with the light from the sun. None are self-luminous. They therefore show phases, as does our moon. If we look over the map, we shall find many things with regard to them. The directions printed with the map on page 79 will explain how it is to be used. The moment or two spent in cutting out slips of paper for the horizons will prove of value.

An Hour Equals 15 Degrees

Mercury, for the month, is near the Sun, which is in the constellations of Sagittarius and Capricornus during the month. If we take a slip of paper and measure on the scale below the map, the distance separating the Sun and Mercury on January 1, 5, 10, etc., we shall find that the distance on the 1st is increased by the 5th. On the 10th it is the same, on the 15th it is less, and it is increasingly less until the 31st. That is, Mercury, at the first of the month, is continually

getting farther away from the Sun until it gets so far that it finally reaches a maximum. This limit, really comes on January 7, and Mercury is said to be at its "greatest elongation west" on that date. It is then 23° 13' from the Sun. (An hour is equal to 15°, as will be easily remembered when we note that twenty-four hours takes us a complete circle of the sky, or 360°.) Mercury, being to the west of the Sun, rises and sets before it. It can therefore be seen only in the morning, and it is called

Cut out slips of paper for the horizons for use with the map

a morning star for this month. It is so close to the Sun as to be hard to see; but if looked for in the proper part of the sky, about January 7, it can be seen rising shortly before the Sun. The sign for Mercury resembles the one for Venus, but the little cup on top reminds one that it is intended for Mercury. This, of course, is only an aid to memory. The sign for Mercury is one that has come down from the ages and has no such meaning.

The Evening Star

Venus, throughout the month, is to the east of the Sun, and so sets and rises after it. It can therefore be seen in the early evening. It is, this month, an evening star. If tested, as was Mercury, it will be found to be continually increasing its distance eastward from the Sun. It is readily seen in the glow of sunset.

Mars—with the sign of the war-like arrow, darting forward over the small circle—will be found to be nearing the Sun. It can also be seen in the early evening, setting after Venus.

Jupiter, strange to say, is not moving to the east as are the others. It is apparently moving *westward* among the stars. It does not move very far in a month, and, being near Gemini, adds greatly to the beauty of the evening sky.

It is almost exactly on the ecliptic, and so serves to show the position of that line in the sky at the present time. On January 1 it is directly opposite the Sun in the sky, and is then nearest the earth for its present circuit. It is then said to be in "opposition" to the Sun. If one is used to judging the daylight hours by the sun, the night hours can be similarly judged by Jupiter around the first of the month.

Saturn is very close to Regulus. It, too, moves westward and but a short distance for the month. It makes a striking showing about the first of the month, and when it is a little above Regulus.

Neptune is too faint to be seen without a telescope. It is also moving westward, and moves only a short distance in January.

But, though the three planets that happen to lie in the upper map all move westward, and all move but a short distance, it is not because they are all in that half of the map, or because they all move slowly, that they move west. The real reason for their "retrograding," or going backward, is best shown by a diagram, and will be explained more readily when one of them pauses just before going in its proper direction eastward.

Uranus is near Mars and is too faint to be seen with the naked eye.

The Moon This Month

The moon at the first of the month is just about to enter Sagittarius. The Moon's positions are for midnight, preceding the date given, in Eastern Standard time. It is west of the Sun on the 1st of the month. On the 2d it is still west of the Sun's position for the second, but on the 3d it is *east* of the Sun. It is easy to estimate that it was at the same position—or right ascension, as it is called—as the Sun at about three hours on January 2. It is then said to be in "conjunction" with the Sun. It is, however, above the Sun, and so does not eclipse it. So on January 2 the moon is a dark ball just above the Sun. We then have "new Moon."

On January 15 it is west of the position of the Earth's "shadow." On the 16th it is still west. On the 17th, however, it is *east*. It is easy to deduce that it is even with the "shadow" at about four hours on the 16th; that is, directly opposite the Sun—i.e., in "opposition," as was Jupiter on January 1. It is now "full Moon."

Since it is below the Earth's "shadow," it is not shaded, and we get no eclipse of the Moon. It is easy to find the date when the Moon is exactly six hours from the Sun and to the east of it. We then have "first quarter." When it is the same distance to the west we have "third quarter." The former position is reached at about six hours on January 9, and the latter at eleven hours and twenty-two minutes on January 23.

The POPULAR SCIENCE MONTHLY intends to publish every month an article by Mr. Hodgson like this—an article that will help you to find your way among the stars. If you feel that the articles can be improved in some way that will be more helpful, write to the editor. And if there are questions that you would like to ask about stars (even though the questions have nothing to do with these articles), send them along.

How to Use the Planet Map

IF we suppose the earth to be a mere dot at the center of a celestial sphere of infinite radius upon which the stars are placed, and if we suppose further that the axis of the earth, produced to meet this sphere, defines its poles, then the celestial equator will lie about the sphere 90° from its poles. The sun in its apparent yearly motion travels along a path called the ecliptic, which is a circle in the sky, tilted with respect to the celestial equator at an angle of about 23½°. The planets and our moon travel in apparent paths that lie close to this ecliptic.

The planet map shows a belt about the celestial sphere, 40° on each side of the celestial equator. It is divided into two halves, so arranged that any point in one half is opposite in the sky to a point immediately opposite in the other half of the map. If we wish the *exact* opposite point, it will be as far above the equator as the first point was below, or *vice versa*. This is easily seen if we imagine the map cut out and joined into a cylinder and set on end. The earth will now be a mere dot in the exact center of this cylinder. If the sun lies at a certain point in one half of the map, the earth's "shadow" must lie in the map opposite, at a point the same distance from the equator, but on the opposite side of it.

The vertical lines are "hour" lines. That is, the earth in its daily rotation turns us about through the stars above us, an amount indicated by the space between two successive hour lines. The stars appear to travel across our southern sky from left to right as this map is drawn.

The lines marked "Eastern Horizon" and "Western Horizon" show how the horizon line will cut the map for the latitude of New York. It is approximately the same for any place in the United States, southern Canada, northern Mexico, and all corresponding latitudes throughout the world. It will be convenient for the reader to cut out two slips of paper having the curve of the western horizon on the left edge of one and the curve of the eastern horizon on the right edge of the other. The points where the curve cuts the parallels should be indicated on the paper, for convenience in placing them. The horizon lines shown indicate the approximate position at 6 P. M. on the first of the current month.

The western horizon is so placed that it cuts the equator immediately above the sun's position for the date in question. The eastern horizon is immediately opposite in the other half of the map. The west and east points are situated where the horizon cuts the celestial equator. They are, of course, twelve hours apart. The zenith, or point overhead, is, for any given position of the horizon, midway between the east and west points and at the extreme top of the map. The south point is 90° below the given zenith which results in its being 10° below the limit of the map.

To determine what part of the sky is visible at any chosen time of any given date, one has only to place the *western* horizon as follows: for 6 P. M. so that it crosses the equator, immediately above the sun's position, *for that date*. For one hour *earlier* (5 P. M.) slip the horizon, parallel to itself, one hour *to the right*; for two hours earlier (4 P. M.) two hours, etc. For one hour *later* (7 P. M.) slip the horizon parallel to itself one hour *to the left*; for two hours later (8 P. M.) two hours, etc. (Note that when one comes to either end of the map, he may proceed to the opposite end of the other and continue as before). Once this western horizon is placed, the eastern is placed in the other half of the map, immediately opposite. Then all the sky to the left of the western and to the right of the eastern horizon is visible at that time on that date.

One can thus tell whether any planet is visible at any particular time. It can also be seen about how far up it is in the sky. Its exact position can then be determined from the adjacent stars.

Mercury lies so close to the sun as to be scarcely ever seen. Venus, Mars, Jupiter, and Saturn can be seen as bright stars, when properly placed with respect to the sun. Uranus and Neptune are too faint to be seen with the naked eye, but are visible in a small telescope. The moon positions are indicated by dots.

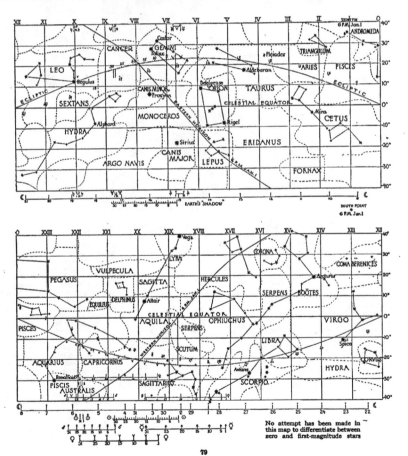

No attempt has been made in this map to differentiate between zero and first-magnitude stars.

Japan Rubs Out "Made in Germany" Mark

"Made in Nippon" is rapidly replacing the old Hun boast. Straw hats and lamp-shades for our trade are made in the delightful studio shown above

This might be a corner "in the garden of my dreams in Tokio"; but a glance at the packing-box in the background shows that the dreaming is of a practical sort. The box will go to California packed with delicate dishes and enticing toys, all with the "made in Japan" stamp on them

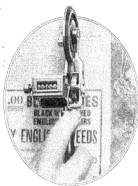

Japanese artisans don't worry much about working clothes; only the boss wears a mustache and a kimono. These men are putting the finishing touches to vases and jars

A Measuring-Wheel to Compute Advertising Space

A TIME-SAVING invention of great practical value to the newspaper and magazine advertising business has recently been patented by Alfred K. Washburn, of Providence, R. I. It is a machine for measuring advertising space in the columns of newspapers or magazines by inches or agate lines and fractions. Heretofore the measuring has been done with a ruler giving on one edge the inches and fractions of inches, on the other edge the agate lines. Agate type is used as the standard for measuring advertising matter.

With the old system of measuring, errors were frequent, and the measuring and adding up consumed a great deal of time. After several trials Mr Washburn, who is a practical mechanic, constructed a machine that reduces the time required for the measuring to a minimum, and not only measures correctly, but automatically adds up the space measured to a fraction of an inch or agate line.

There is no "human equation" in this little wheel, which measures type space quickly and accurately

The inventor made use of the same principle on which the measuring wheels of surveyors and draftsmen are based, but he adapted the idea to the specific needs and requirements of the advertising business, and introduced a number of particularly useful improvements, calculated to save as much time and labor as possible without diminishing the exactness of the measurements.

The measuring in Mr. Washburn's invention is done with a wheel, which is rolled over the matter to be measured. It revolves between the two prongs of a fork with a conveniently shaped handle. It can turn in only one direction.

By a gear the wheel is connected with the counting device, which records each complete revolution of the wheel. An indicator on the wheel itself shows on a graduated scale the distance covered, which is not recorded by the counting device because it is less than the circumference of the measuring wheel.

A Dictaphone Made from an Old Phonograph

By E. F. Hallock

THE modern dictating machine differs from the old cylinder phonograph chiefly in the following respects: It is driven by an electric instead of a spring motor so that winding is done away with; it is fitted with a slightly larger cylinder, both as to diameter and length, and with a finer worm feed so that it takes eight minutes to fill up—or listen to—a cylinder, as against four minutes for the phonograph and, of course, the transmitting and reproducing style are rather cut the better to suit them for the finer track which the slower feed gives. The dictating machine is fitted with a device which raises the recorder or the reproducer from the record when the flow of speech is interrupted, or when it is desired to

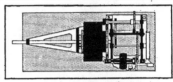

The solenoid and where it is placed to work the arm attached to the reproducing carriage

halt its reproduction for an instant or in order to put the words into type, and this control is effected by means of a pedal device; a celluloid scale is provided so that the dictator can, with each cylinder, make a notation of repetitions, parts to be left out or changed, etc., for the guidance of the transcriber.

The most formidable of these differences to overcome is the length of time which it takes to complete a record. Upon investigation, however, it was found that the earliest dictating machines were equipped for five-minute recording or reproduction, and accordingly a five-minute recorder was purchased and the sapphire style of reproducer was changed for the finer five-minute type. This equipment was tried out thoroughly before the mechanism of the phonograph was altered and it was found that by

decreasing the speed of the machine slightly, a full five minutes can be gotten out of each cylinder despite the coarser worm; nor does either the faster speed or the slower cylinder rotation sacrifice aught in the clarity of the reproduction.

The next point considered was the means of raising the reproducer arm to lift the stylus from the record and at the same time stop the feed, or forward movement, of the reproducer carriage. On the phonograph, this is done by means of a small hand lever, a method obviously too inconvenient where one needs both hands to operate the typewriter. The usual method on the commercial dictating machines is to use a pneumatic control similar to that used to snap the shutter on a camera; but, unless carefully constructed, such a device is apt to leak air and give rise to endless trouble. It was abandoned, therefore, as impractical where tools for accurate workmanship were lacking, a simple electrical control being deemed easier to construct and less likely to become deranged.

The basis of the electric control device is a solenoid shown in the illustration Fig. 1. It consists simply of a 5-in. length of brass tube, ⅝-in. internal diameter, around which is wound 1 lb. of ordinary double cotton-covered annunciator wire, No. 18 gage, to form a coil about 3½ in. in length between two fiber disks driven to a tight fit on the tube. Its armature consists of a piece of ⅜-in. iron gas pipe 2 in. long, drilled at the top as a means of attaching a small "eye" of copper wire to which a length of silk fishing line is attached.

This simple solenoid is placed on the floor behind the table or desk on which the dictating machine rests, and transmits its motion to an arm attached to the reproducer carriage at the same point as the feed nut spring, as shown in Fig. 2. This arm is a piece of aluminum, ⅜ in. wide, ¼ in. thick and 8 in. long, which is drilled for the passage of the screws used to hold the feed nut spring. The original screws are replaced by longer ones to account for the thickness of

the arm. The outer end of the arm has 3 holes drilled for the silk cord to pass through, as shown, providing a ready means of adjusting the length of the cord in case the machine is moved to a new location.

The solenoid is located so that its axis is in line with the midpoint of the

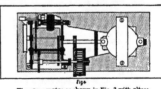

Fig. 3.—The layout of the original spring motor as it was used to drive the phonograph record

feed worm and also with the outermost hole drilled for the passage of the cord in the extension arm, and the length of the cord is adjusted so that the top of the armature is just flush with the top of the brass tube of the solenoid as shown. The solenoid takes its current from a small transformer which steps down the house lighting current to 9 volts and which is located in the cabinet of the phonograph. These transformers can be purchased for a nominal sum for operating bells and small electrical toys. A foot push-button is placed in the circuit with sufficient length of flexible cord to bring button to a position for convenient use by the operator.

In Fig. 2 is shown the celluloid scale and the pointer. The latter is a small piece of strip brass riveted to the end of the feed nut spring with the same rivet which holds the feed nut to the spring, it being necessary to replace the original rivet with a new one, of course. The pointer is bent

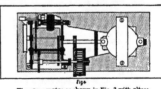

Fig. 4

The same motor as shown in Fig. 3 with alterations to adapt it for electric operation

to bring the point down close to the scale when the reproducer is brought into action. A couple of pieces of 28-gage brass cut in semi-circular shape are screwed to the base plate of the machine to retain the scale; their straight edges are bent up slightly from the plate to facilitate removal and replacement of the scale, which is somewhat slightly longer than the full travel of the feed nut on the worm.

The original plan was to discard the spring motor altogether and drive the machine by means of a small electric motor belted directly to the pulley on the cylinder mandrel. The almost certainty of trouble arising from speed fluctuations and consequent distortion of the speech, however, led to that plan being abandoned for one wherein the spring of the original motor is simply replaced by an electric motor, the friction speed governor of the spring motor being retained intact.

The illustration Fig. 3 shows in detail the layout of the spring motor before alteration, and Fig. 4 shows the alterations made to adapt it for electric operation. The entire spring case with the winding shaft, gearing, ratchet and pawl, and also the main drive gear, were removed and discarded. The secondary shaft, which was directly driven by the engagement of its pinion with the main drive gear, was taken out, the pinion removed and discarded and the gear removed. This gear was placed in relatively the same position on an arbor of the same diameter, which, however, was $2\frac{1}{4}$ in. longer.

On the extension of this shaft was mounted a new spring case, the basis of which is a gear wheel $3\frac{1}{2}$ in. in diameter, the largest which can be used without cutting a pocket out of the cabinet. The case itself is a section of brass tube 3 in. in diameter and $\frac{3}{4}$ in. long soldered to the gear wheel and provided with a brass cover soldered in place after the spring is fitted. The spring is of steel, of course, $\frac{1}{2}$ in. wide and 2 ft. long and not too stiff. It is riveted to the case so that its spirals are in an anticlockwise direction. The inner end of the spring fits into a saw slot cut in the shaft so that the whole casing can be slipped into place after the cover has been soldered down. Both the gear and the casing cover, of course, are a loose fit on the shaft so that they can rotate independently of the shaft; the casing is prevented from slipping off the shaft by means of a brass collar soldered to the shaft on the end, movement in the other direction being prevented by the spring itself coming up against the end of the saw slot.

The electric motor was taken from a discarded medical vibrator and operates on 110 volts. The armature shaft as well as the vibratory mechanism was removed and discarded, the shaft being replaced by a longer one of the same diameter. The bearing holes which served for mounting the original spring were reamed to take the electric motor shaft, and the motor was clamped in place between two strips of $\frac{1}{8}$-in. steel near the apex of the triangular cast iron frame of the original motor. Filler blocks of rubber made from sections of an old automobile tire were used between the motor and the clamp plates to provide the

slight adjustment necessary to bring the shaft into line with the bearings on the base plate and also to help muffle any noise due to the operation of the motor.

The motor was arranged to rotate in a clockwise direction and on its shaft a pinion $1\frac{1}{4}$ in. in diameter was fitted in a position to engage with the gear on the new spring housing when the motor was clamped in proper position.

It will at once be apparent that the motor speed is very slow, a fact which conduces to silent operation and absence of vibration both of which are essential if good results are to be obtained in the reproduction of the spoken words. Any unevenness of torque due to the slow speed of the motor and any tendency to jerkiness resulting therefrom are absorbed by the spring drive, the spring being kept constantly wound to its proper operating tension by the motor. While the slow speed of the motor does not make for cool operation, it has not been found in months of actual usage, that the heat approaches the dangerous point.

The electrical connections, of course, are simple. The motor is directly connected to the house mains by means of a special connection plug while the primary of the solenoid is connected in multiple with the motor, so that the operation of a snap switch located on the side of the cabinet turns the current on or off for both circuits.

It is important, in fitting the feed nut extension arm, to replace the feed nut spring so that the threads of the nut engage properly with those of the feed worm; and important also to make sure that the tension of the spring is sufficient to insure the regular travel of the worm under the slightly additional weight. It may be found necessary, in some cases, to increase the tension by bending the spring down toward the worm slightly, and where such is the case it probably will be necessary to counterweight the weight of the solenoid armature and the extension arm by placing a lead washer of proper weight around the orifice on the reproducer.

Cleaning Grease-Filled Surfaces on Emery-Wheels

WHEN an emery-wheel becomes coated with grease it fails to give satisfactory results. Emery that is to be cleaned should be washed with dilute acid after boiling it with a solution of caustic soda. Carbon bisulphide should be used.

Holding One End of a Mower Cutter while Grinding

TO relieve the weight on the arm while grinding a mower cutter, fasten a rope to the head end of the bar and pass the rope over a pulley suspended above. Bring the rope down to the hand holding the knife-bar. This will hold the end up easily.

As you change from one knife to another in grinding, the rope in the hand can be let out or taken up as required for the proper angle. If the tool-grinder is portable, the pulley is not a necessity, since the rope can be passed over a tree limb or any horizontal stick which is overhead.—GLENN A. GRANGER.

The end of the bar is held up on a level with the grindstone

A New Sign-Post Made of Metal Tubes

THE old-fashioned sign-board with the pointing hand is still used ordinarily for directing the wayfarer to points of interest; but, at best, it affords a poor line of sight, and soon succumbs to the elements.

The accompanying illustration shows a simple and durable sign-post. The sighting is done through a piece of 2-in. pipe about 20 in. long, which is supported on top of a piece of the same size pipe set firmly in the ground and standing up about 4 ft. A piece of $\frac{1}{8}$-in. sheet iron is bent double, fitted over the sight-

The set tube pointed toward a mountain shows the traveler what he is viewing

ing pipe, and riveted at each end. Connection with the vertical pipe is made by running a 5/16-in. bolt through from one side to the other. A second bolt is then placed lower down, and operates in a circular slot so as to permit the line of sight to be adjusted to the proper vertical angle, after which the nuts are adjusted tightly. The name of the city should be painted on each side.

A Comfortable Chair Made From a Twelve-Inch Board

SEVERAL chairs designed by a local artist for his studio are quite remarkable in appearance, besides being comfortable seats. They were made almost entirely from a 12-in. board.

There are only four parts to each chair: One 12-in. board 39 in. long, which is both back and front "legs."

One piece 14 in. long, which makes the seat; and two triangular pieces of the same material and length serving the double purpose of rear legs and supports for the seat.

Studio chairs made from a single board cut to make the seat and shape the rear legs

The main piece is inclined at an angle of about 40 deg. to the floor and the braces are so cut that their greatest, upper, angle of 130 deg. is divided by the main piece. In other words, both the seat and the rear edge of the braces are at an angle of 65 deg. to the main piece.

The brace-pieces are mortised into the main piece and fastened with screws, and the seat is also screwed to the braces and to the back.

It is probable that a simpler chair deserving the name could not be made. On the other hand it is really astonishing how much comfort is to be had from these narrow and angular pieces of furniture. This is due to the angle of inclination of the back, which permits a position of the body possible only in a rocker or a Morris chair—or by tilting back an ordinary chair.

These chairs are cheap, quickly made, strong and durable and could be used to advantage in a good many places where uncomfortable seats would cost just as much or more.

A Home-Made Prick Point for the Draftsman

THE instrument shown was made from a rubber pen-holder with a wood plug cut and tapered to put

4"Of hard rubber penholder   1"Hard wood   Needle

An old rubber penholder makes the handle for a prick point fitted in a plug

in the hole. One end of the wood plug is fitted with a needle. When not in use the plug is reversed and stuck into the tapered hole of the penholder.—THOMAS O. WANSLEBEN.

A Draining Rack and Press for Making Cheese

WHEN the Dairy Division of the United States Department of Agriculture is conducting a nation-wide campaign to introduce methods of manufacturing and use of cottage cheese on every farm, a cheese-making equipment as here described is simple and can be made at home. It can serve profitably in the capacity of converting surplus milk into a valuable food in the family diet. There are approximately 10,755,790 farm families in the United States and the production of two pounds of cottage cheese a week would aggregate 1,118,603,160 pounds yearly.

The rack for draining the cheese is 16 in. deep, 12 in. wide, and 24 in. long, and is cut from pine timber. The bottom slats which hold the pan under the draining cloth fit into notches made in the lower side strips and can easily be removed when the rack is washed. The corner posts extend ¾ in. above the strips at the top and the corner

A small wood frame to support the cloth for draining curdle to make cottage cheese

loops of the muslin or cheesecloth used as a drain cloth are looped over the posts. A rack can be made out of an orange or vegetable crate as shown in the accompanying illustration.

The press consists of two poplar or maple boards 1¼ in. thick and 14½ in. square. Strips of wood 1¾ in. wide are nailed or screwed on the back of each board to prevent them from warping. The boards are planed and sand-papered until altogether smooth. The lower board has a circular groove which has an outlet through which the whey drains as it is pressed out of the curd.

A wooden paddle, a dairy thermometer, and a food chopper or sausage grinder complete the cheese-making outfit for farm use. The molding tube or cylinder can be made by a tinsmith or purchased from a hardware store.

The paddle can be made easily at home. The molding tube is designed to mold the cheese into attractive and

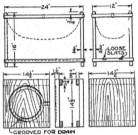

Dimensions of the parts for making the stand and the parts for the press

convenient form for the market. The cheese can also be packed into small glass jars by placing the opening of the jar over the end of the tube through which the cheese is forced.—S. R. WINTERS.

A Solution for Removing Grease from Machinery Parts

THE following method has been substituted for the use of gasoline and other light oils: Boil the parts in caustic soda lye,—1 lb. to a gallon of water,—then brush while the articles are hot. Caustic soda is recommended as better than ordinary soda, since it causes the fat or grease to dissolve more quickly.

How to Raise a Stuck Window in a Railway Coach

THE windows of the average railway car very frequently stick or jam when closed, and cannot be opened no matter how strong a lift is applied to the lifting lug. When once raised even a little, the window invariably rises the rest of the way without any trouble.

The sketch shows the shank of the foot-rest of the seat usually found under the seat ahead in a position for a handy lever to loosen the window. By getting the end of the slide under the finger lug and resting the remaining portion on the sill, a leverage is readily exerted which will loosen any obstinate window. The foot-rest and slide slip very easily out of their slot under the chair.

The foot-rest end placed under catch

This discovery may be of interest to some suffocating traveler sitting near a closed window.—F. W. BENTLEY.

An Electroscope for Detecting Small Electric Discharges

A SENSITIVE electroscope is always a most useful if not an indispensable instrument in the equipment of every laboratory. Use has generally been made of the electroscope in detecting and determining the nature of minute electrical charges. The same principal, however, has been applied to a method of measuring delicate potentials and, quite recently, to the detection of feeble radio-active bodies. Since the former application may more properly be termed an electrometer and the latter involves an apparatus beyond our immediate needs, a very sensitive type of the common gold-leaf electroscope will be the subject of this article.

The flask of the electroscope has a capacity of about twenty-one cubic inches and a mouth of about 1⅝ in. in diameter. Such a flask can be purchased for a small sum from some chemical supply house. Its height should be about 6¾ in. from its base and be about 3¾ in. in diameter at its largest point. Two ¾-in. brass balls should be procured at the same time.

When these have been obtained proceed as follows: Drill a ⅛-in. hole to a depth of 3/16 in. in one of the brass balls and saw the other in half. One of the hemispheres may be discarded but the other must have a ⅛-in. hole drilled through its center perpendicular to its plane face. Tap both of these holes with a 10/32 machine thread. Care should be taken to protect the balls during the drilling and tapping so that their surfaces will not be marred. A piece of 3/16-in. brass rod must be cut so that there are two lengths; one 2¼ in. long and one 3¼ in. long. The shorter of these two rods must have both its ends threaded to a distance of 3/16 in. with a male thread corresponding to that in the ball and hemisphere. Thread one end of the longer rod similarly, but with the aid of a hack-saw cut its opposite extremity lengthwise to a distance of ¼ in. Cut a rectangular piece of 1/16-in. brass, ½ by ¾ in., and insert the same longitudinally in the slit prepared in the brass rod and solder fast. After the superfluous solder has been removed with a file, the rectangular plate may be trimmed as shown at *A*.

Screw the ball, *B*, to one end of the short rod, *C*, and the hemisphere, *D*, plane face out, to its other extremity. After the long rod, *E*, has been screwed to the plane face of the hemisphere, polish the brass with pumice-stone and lacquer everything except the ball, *B*, and the plate, *A*.

Obtain a cork somewhat larger than the mouth of the flask and with the help of a sharp razor blade trim it so as to make a nice fit, care being taken not to have the fit too tight for fear of breaking the thin walled flask. Place the cork on the floor and roll under the foot until it has become a

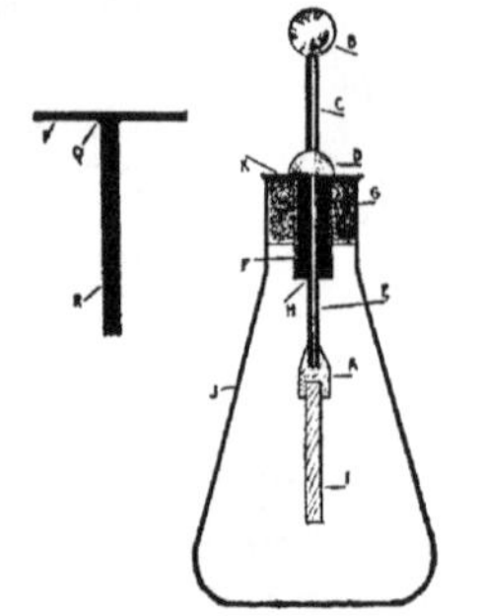

The flask and its mountings for holding the gold foil, which is the most delicate part

great deal softer. Sandpaper its sides smooth and bore a ⅝-in. hole through its center with a cork-borer. Cut the cork to a length of 1 in. A piece of glass tubing, *F*, with a ½-in. bore and a 1/16-in. wall must be cut so that it is 1⅝ in. long. This is then inserted in the hole prepared for it in the cork, *G*, in such a manner that one end projects ⅛ in. above the top

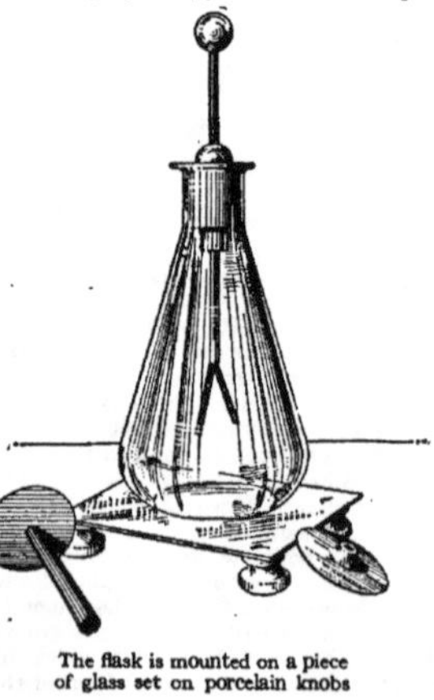

The flask is mounted on a piece of glass set on porcelain knobs

of the cork and the other end falls ½ in. lower than its bottom. Place the long rod, *E*, through the glass tube, *F*, so that the plane face of the hemisphere, *D*, shall rest upon that end of the tube projecting ⅛ in. above the cork. Invert the whole and fill the glass tube with molten sealing-wax, *H*. When the insulation has cooled the rod will be embedded in its center.

Some gold-foil, which can be purchased for a small sum from a glazier, must be cut into strips ¼ in. wide and 2¼ in. long. Try to have this cutting done by a professional, but if this is impossible it is best done by leaving it between the sheets of paper as it comes and cutting paper, foil, and all. The plate, *A*, should be smeared with albumen, thin shellac, or, better still, a solution made by dissolving a large capsule in a teaspoon of warm ether, as an adhesive. Two strips of the gold-foil, *I*, are then affixed to opposite sides of the plate, leaving 2 in. of the strips hanging free. Fixing and attaching the gold-leaves is the most difficult operation in the whole process and much patience will be required before two leaves can be properly fastened. The foil, of course, cannot be touched by the hand and must be handled exclusively by some improvised instrument such as a narrow strip of paper. Few hints can be given as to the best way to proceed with the attachment, so the constructor must rely upon his own resources and go about the task as an ardent supporter of the ancient maxim that "experience is the best teacher."

The flask, *J*, must be clean and dry. Before placing the cork containing the rod into place permanently, heat the flask for a few minutes at a moderate temperature so as to drive all the moisture from the interior. While the flask is yet warm, place the cork in the top so that it is ⅛ in. below the mouth. Pour molten sealing-wax, *K*, on the cork until it is level with the top. Give the neck of the flask several coats of shellac on the outside to prevent moisture approaching the insulation.

If it is desirable to make a condensing equipment for the electroscope, cut two disks, each 2½ in. in diameter, from 1/16-in. brass. By means of solder, *L*, affix a piece of ⅜-in. brass rod 5/16 in. long, *M*, to the center of one of the disks, *N*. Drill a ⅛-in. hole, *O*, in the end of the rod to a depth of 3/16 in. and tap as was the case with the charging knob. This constitutes the condensing plate which, when in use, displaces the charging ball.

The other disk, *P*, which forms the collecting plate is fastened by means of wax, *Q*, to the end of a rod, *R*, of some insulating material such as hard rubber, glass, or fiber, 3½ in. long. Polish both plates and lacquer. The lacquer must not be dispensed with for it is necessary as the dielectric of the condenser.—MARK M. GEATRY.

You've got to have the right pencil

If your pencil work is to be always up to top speed, top efficiency, you've got to have a pencil not only of the best quality but of the degree of lead exactly suited to your work.

DIXON'S ELDORADO
"the master drawing pencil"

is proclaimed by artists, architects, engineers, business men and other connoisseurs, as a real American achievement. The strong, long-wearing leads do not easily break or wear down quickly. Their responsiveness makes your work less tiring and quicker. First of all specify the Eldorado; then be sure you are getting the right grade.

How to find your grade. Note in the chart below that 6B is the softest, 9H the hardest and HB is medium (the degree most used in general work). Select the degree you think will suit your work and your liking. If not exactly right, next time choose a grade harder or softer. When you have thus found *your* degree, specify it every time and you will always have pencil satisfaction.

Send us 16c in stamps now, specifying the degrees you want and we will send you full-length samples worth double the money.

JOSEPH DIXON CRUCIBLE CO.
DEPT. 120-J   JERSEY CITY, N. J.

Canadian distributors Established 1827
A. R. MacDougall & Co., Ltd., Toronto, Ont.

6B	Varying degrees of	H	Hard.
5B	extra softness — 6B	2H	Harder.
4B	softest.	3H	Very hard.
3B	Extra soft and black.	4H	Extra hard.
2B	Very soft and black.	5H	Varying
B	Soft and black.	6H	degrees of
HB	Medium soft.	7H	extra
F	Firm.	8H	hardness.
		9H	

A Removable Bag Receptacle for a Park Rubbish-Container

A DURABLE and inexpensive depository for scraps of paper and rubbish that accumulate around re-

The cloth bag is held open by a metal ring and it can be easily removed for dumping

sorts and picnic grounds may be made in the following manner:

An iron ring large enough to stretch a gunny-sack over is provided with six projecting hooks riveted firmly in place. This ring is then riveted to the top of a 2-in. pipe, the connection being strengthened by the addition of a piece of ½-in. steel plate curved to the same radius as the ring, and riveted to both the ring and the vertical pipe, which is then set firmly in the ground, preferably in a cement base.

After painting and lettering the plate, drop a gunny-sack through the ring and stretch the upper edge over the hooks. Such a receptacle may be easily emptied, and it is a very simple matter to replace the sack when it is worn.—JOHN D. ADAMS.

Sights for More Accurate Shooting with a Shot-Gun

IT has always been understood among small-game hunters that a crack rifle shot could not hit anything with a shot-gun, in spite of the fact that his aim with this firearm does not have to be very accurate. Here is the way I fitted out my shot-gun so that the

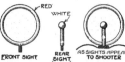

FRONT SIGHT   REAR SIGHT   AS SIGHTS APPEAR TO SHOOTER

The ring with a white ball center makes it easy for the hunter to sight his gun

aim is just as precise and definite as that of any rifle.

A ring was mounted on the screw for the front sight, the ring being 1¼ in. in diameter (inside measurement). The rim appears to the one sighting

---

the gun to be about ⅛ in. thick. The ring was painted a bright red, as red shows up best against any kind of a target and stands out very distinctly when aiming at moving game.

A small hole was drilled and threads cut to receive the ring-screw 2 in. from the muzzle end of the barrel. The rear sight is a vertical post ⅛ in. in diameter, topped with a ball ¼ in. in diameter, all projecting ¾ in. above the barrel. This is painted a dull white color. The post screws in place 24 in. back of the front sight. With this combination of white rear post and red ring I find that I can easily bag my share of game. It is also a very clever sight for the "rifle crank" to use when he shoots the scatter-gun at traps. The principle involved in this kind of sight is this: when the white ball aligns anywhere within the ring with your target you get your game.—F. E. BRIMMER.

Centering Round Stock with a Vise and Scriber

A VERY simple method of finding the center on a round piece of iron or a bar is to clamp a center punch or scriber in the jaws of a vise, so that its point extends above the jaws to about the radius of the bar to be marked. The bar is placed on the

To get the center of a round bar, roll it between the jaws of a vise holding a scriber

jaws and revolved by hand against the scriber point.

If the point is not exactly in the center, a circle will be marked so that its center will be easily found. The fact that the jaws are slightly open on account of the scriber being clamped in them provides a trough, or ways, in which the bar may be easily rolled.—JOHN SCHMELZEIS.

How to Make a Nickel Alloy that Is Malleable

MAGNESIUM added to nickel or cobalt, when put into the fused metals in the proportion of one eighth of one per cent, will render these metals malleable. The nickel will also become ductile when mixed with this small proportion of magnesium, while the cobalt loses its color and becomes whiter than the nickel. Both of these metals can be made to adhere firmly to iron or steel at a white heat.

A Novel Rope-Holding Device for the Ground

A LIGHT form of a deadman that is very convenient for a pull directly upward is shown in the accompanying illustration. It consists of

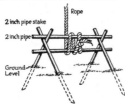

With the cross stakes at each end and the two parallel bars as shown the whole cannot be drawn from the ground

stakes or short pieces of pipe driven in the ground at an angle, with cross-pieces above and below where the stakes cross. The rope is secured about the two cross-pieces.

This form of a deadman is quickly set, and will stand any pull that will not break the stakes or bend the pipes.
—WALTER L. MORRISON.

Removing Dampness and Disinfecting a Cellar

C ELLARS will acquire a musty odor after being closed up for the winter. To remove dampness as well as to disinfect the cellar, sprinkle chloride of lime on the floor and close up the cellar for a few days. Then open the windows and let in the air until the chloride-of-lime odor disappears, and your cellar will be ready for storing vegetables.

How to Make a Valve for a Small Steam-Engine

A SMALL valve for use on a toy steam-engine or an engine for chemical purposes can be easily made in the following manner. Secure a block of brass ⅝ in. square and ¾ in. long, and drill a hole through its center. This hole will be governed by the size of the pipe. The hole should be threaded at each end to receive the pipe ends. Drill another hole on the surface at right angles to the first hole, as shown at A. Make the diameter of this hole ⅓ in. larger than the first hole. Pass a valve consisting of a thumb-screw threaded to fit the threads cut in the last drilled hole through the hole for opening and closing the line.—THEODORE P. GATHMANN.

Small steam-valve made from a piece of brass

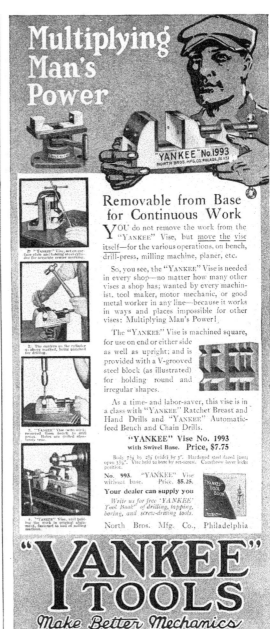

Our Nation's Flags and Their Meaning

STRICTLY speaking, the word "flag" is not used to any great extent in the army. The national flag belonging to each regiment is known as the "national colors," and the regimental flag is the "regimental standard." The company, troop, and battery flags of the engineers, cavalry,

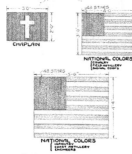

CHAPLAIN

NATIONAL COLORS
CAVALRY
FIELD ARTILLERY
SIGNAL CORPS

NATIONAL COLORS
INFANTRY
COAST ARTILLERY
ENGINEERS

Chaplain's flag and regimental flags, the latter being the national colors

and artillery are known as "guidons"; and so on down the line.

A knowledge of the meaning of every flag in use in the army will be found useful not only to the soldier but to the civilian who is, or should be, interested in things military. The accompanying illustrations will be of great assistance in distinguishing various bodies of troops, which will be met more frequently as time goes on, as well as in distinguishing regimental, brigade, and division headquarters.

Most army posts are entitled to

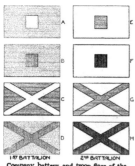

1ST BATTALION   2ND BATTALION

Company, battery, and troop flags of the first and second battalion, known as guidons

three flags—garrison, post, and storm. These are what is known as national flags, and are made of bunting in the sizes indicated. The "garrison flag"

is furnished only to posts designated from time to time by the War Department, and is hoisted only on holidays and important occasions. The post

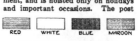

RED   WHITE   BLUE   MAROON

ORANGE   YELLOW   BLACK

How the colors of the various flags in the illustrations are designated

flag is furnished to all army posts, and is always run up in pleasant weather. The storm flag is furnished for all occupied posts, for use in stormy and windy weather. The last named flag is also hoisted in all kinds of weather over semi-permanent camps and national cemeteries.

The colors of the President are of blue silk, and are attached to a single jointed staff 10 ft. 3 in. long, including the ferrule and gold-plated head. The head consists of a globe 2 in. in diameter surmounted by an American

PRESIDENTS COLORS

SECRETARY OF WAR

ASST. SEC. OF WAR

The colors and sizes of the flags of the President, the Secretary of War, and his assistant

eagle, alert, 5⅜ in. high. In each of the four corners is a five-pointed star, white, with one point upward, the points of each star to lie within the circumference of an imaginary circle of .468 ft. diameter; the centers of the circle being .77 ft. from both the long and short sides of the colors. The coat-of-arms is placed in the center of the flag, the letters and stars being embroidered in white silk on both sides. The colors are trimmed on three sides with a knotted fringe of silver and gold 3 in. wide. The cord is 8½ ft. long with two tassels, and is composed of red, white, and blue silk strands.

The flag of the Secretary of War is of scarlet bunting having a five-pointed star in each of the four corners, each

star having one point upward. The points of the star lie within an imaginary circle of 5-in. radius. The centers of the stars are 17 in. from the short side and 12 in. from the long side of the flag. The center of the flag bears the official coat-of-arms of the United States.

The chaplain's flag is of blue bunting having in the center a white Latin cross 18 in. high and of suitable width.

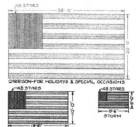

GARRISON-FOR HOLIDAYS & SPECIAL OCCASIONS

POST-FOR PLEASANT WEATHER

STORM

Army posts have three flags—garrison, post, and storm. The garrison flag is only given to certain posts by the War Department

These flags are used for field service.

Hospital flags are of three sizes, and are used in connection with the national flag in time of war. The guidons are used to mark the way to field hospitals.

National colors are of silk, 5½ ft. fly and 4 ft. 4 in. on the staff, which is 9 ft. long, including the spearhead and ferrule. The union is 2½ ft. long, and the stars are embroidered in white silk on both sides. The edges are trimmed with knotted fringe of yellow

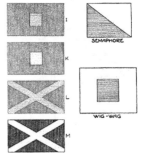

SEMAPHORE

WIG-WAG

3RD BATTALION

The third battalion flags, together with the semaphore and wigwag flags

silk 2½ in. wide. A cord 8½ ft. long is provided, and this cord is made of red, white, and blue silk strands. National colors are practically identical

Aladdin Homes

Make This Your Home Building Year

That charming bungalow Home—bordered in flowers—bathed in sun-light—the one you have dreamed of can now be yours. It may be a snug and cosy cottage complete with five rooms on one floor, or a more pretentious bungalow of charming proportions. These types and many others are found in the Aladdin book—complete home-building library of 100 designs—100 ideal homes. Besides this there is a vital message for you in this most remarkable book, that will be sent at your request.

Saves Lumber and Labor and You'll Save Money Building Your Home

Of course, you would like to save money building your new home—would like to build at the lowest price possible. But, you cannot expect to save very much money by bargaining or getting different prices. You can only expect to save money by saving labor in building your home and eliminating waste of lumber. Aladdin Readi-Cut Houses are cut-to-fit in Aladdin mills by rapid power machines. Old fashioned building waste is reduced from 18% to less than 2%—a saving of practically $18 on every $100 worth of lumber. And you save a third of the labor in building your home. This saving of lumber and labor will mean from $200 to $400 or more. The Aladdin Book shows you how it is done. Send for this book today.

Aladdin System Saves Waste

Thousands of our customers have told us of big savings they made building Aladdin Homes. The average saving on our $945 bungalow is $264. The Aladdin system saves waste of lumber. All of the material is shipped to you cut to fit. Your carpenter can erect your Aladdin Home quicker because all of the hand sawing is done in our mills by Rapid Power Machines.

Aladdin's Dollar-a-Knot Guarantee

Aladdin Readi-Cut houses are the finest produced. The choicest stock available is used for the exposed weather parts. Finer materials cannot be had than those used in the construction of Aladdin Readi-Cut Homes. The Aladdin Dollar-a-Knot Guarantee is but one evidence of the built-in quality which is a part of every Aladdin Readi-Cut Home.

What You Get With Your Aladdin Home.
Aladdin houses are cut-to-fit—no waste of lumber or labor. The Aladdin price includes all materials cut-to-fit as follows: Lumber, millwork, flooring, Outside and inside finish, doors, windows, shingles, lath and plaster, hardware, locks, nails, paint, varnishes. The material is shipped to you in a sealed box car, complete, ready to erect. Safe arrival of the complete material in perfect condition is guaranteed. Send stamps today for a copy of "Aladdin Homes" No. 650.

THE ALADDIN COMPANY, 651 Aladdin Ave., Bay City, Mich.
Canadian Address: Canadian Aladdin Co., Ltd., Toronto, Ontario.

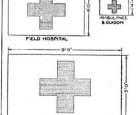

FIELD HOSPITAL
AMBULANCE & GUIDON

GENERAL HOSPITAL.

The size and colors used for the general field hospital and ambulance flags

for engineers, infantry, and coast artillery branches, the designation of the regiment or battalion to which the colors belong being engraved on a silver band placed around the staff. For field artillery, cavalry and signal

FIELD ARMY HEADQUARTERS

INFANTRY DIVISION HDQTS.   CAVALRY DIVISION HDQTS.   ARTILLERY BRIGADE HDQTS.
INFANTRY-BLUE FIELD
CAVALRY-YELLOW FIELD

Field army headquarters, infantry, cavalry and battery brigade flags

corps the national colors are mounted on a lance 9½ ft. long. The union is 22 in. long, the stars being embroidered, and the fringe being the

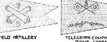

CAVALRY

AERO SQUADRON (SIGNAL CORPS)

FIELD ARTILLERY

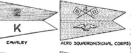

TELEGRAPH COMPANY (SIGNAL CORPS)

ENGINEERS

FIELD HOSPITAL

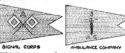

SIGNAL CORPS

AMBULANCE COMPANY

Guidons are used to identify troop, battery, etc., from one another and so aids marching troops

same in color as other national colors. The regimental standards are of the same dimensions as the national colors, but vary as follows:

	Field	Fringe	Color of Scroll	Color of Inscription
Engineers	scarlet	white	white	white
Mounted Engineers	scarlet	white	white	white
Coast Artillery	scarlet	yellow	yellow	red
Infantry	blue	yellow	red	white
Cavalry	yellow	yellow	red	yellow
Field Artillery	scarlet	yellow	yellow	scarlet
Signal Corps	orange	white		

The inscriptions on the regimental standards are as follows:

U. S. Engineers
Battalion, U. S. Mounted Engineers
U. S. Coast Artillery Corps
U. S. Infantry.
U. S. Cavalry
U. S. Field Artillery
Battalion Signal Corps, U. S. A.

Guidons are used to identify troops, batteries, etc., from one another, and also to help guide troops on the march. The guidons for the various arms of service are shown in the illustration.

AMMUNITION TRAINS, DISTRIBUTING POINTS AND DEPOTS

These guidons are made of bunting for field service, and of silk for battles, parades, and reviews.

The infantry

TELEGRAPH STATION

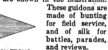

MOTOR TRUCK COMPANY

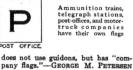

POST OFFICE.

Ammunition trains, telegraph stations, post-offices, and motor-truck companies have their own flags

does not use guidons, but has "company flags."—GEORGE M. PETERSEN

To Prevent Unconscious Retarding of the Automobile Spark

HAS any Ford driver, who was too tall to fit his car, ever noticed that, when driving at 20 miles an hour or more with the spark well down, his knee is always pushing the lever up again, and so he is continuously though unconsciously retarding the spark?

This difficulty may be overcome by bending the lever 3 in. more than the original bend. It is astonishing what a difference will be found in the available space.

To bend the lever, however, is more easily said than done; it is necessary to use two large adjustable wrenches. One should be slipped over the short bend of the lever to hold it in position, and the other should be used to do the bending. The second wrench should be placed at the extreme end of the lever, and in some cases it may be necessary to slip a piece of gas-pipe over the handle to obtain sufficient leverage. Get someone to hold the first wrench for you while you do the bending.—J. S. CHAPMAN.

GOODELL-PRATT

1500 GOOD TOOLS

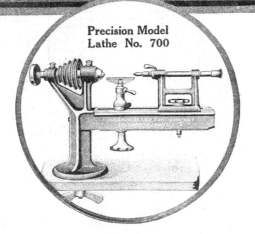

Precision Model
Lathe No. 700

A COMPLETE lathe designed for the accurate tooling of small work.

This lathe has a 12-inch bed and an extreme distance of 3½ inches between centers. It swings 5 inches. This lathe is provided with a draw-in spindle with a ⅜-inch hole clear through. The pulley has four steps for ¼-inch round belt. The compound slide rest, boring attachment, sawing attachment and milling attachment, shown in the four circles below, are intended for use with Precision Model Lathe No. 700. With these attachments an almost unlimited scope of work can be accurately and quickly accom-.

plished. This lathe is particularly adapted to small precision work such as is usually done in the laboratory. Doctors, dentists, jewelers and inventors will find it exceedingly useful in their laboratories.

Into this lathe, just as into all of the 1500 Goodell-Pratt tools, is built quality. Nothing in workmanship or material is spared which will in any way improve Goodell-Pratt tools. The policy of designing each tool to fulfil a definite purpose, and the determination to make every tool the best of its kind, has resulted in mechanics the world over demanding Goodell-Pratt Tools.

GOODELL-PRATT COMPANY, *Toolsmiths,* Greenfield, Mass.

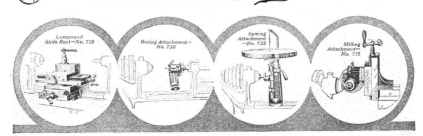

Compound Slide Rest—No. 710

Boring Attachment—No. 720

Sawing Attachment—No. 725

Milling Attachment—No. 715

"The Construction

of Small

Alternating Current

Motors"

By A. E. Watson, E. E., Professor of Electrical Engineering in Brown University, Providence, R. I.

This book contains complete instructions for building small alternating current motors in several sizes. The designs will be found to be in harmony with those of the very best manufacturers.

Important Information for Electricians

Some of the subjects taken up are "Characteristic features of alternating current motors," "Construction of a one-half horse-power, single-phase induction motor," "Construction of a one-kilowatt, two-phase or three-phase alternating current generator or a one horse-power synchronous motor," "Procedure in testing and using an alternating current generator or synchronous motor," "Construction of a one-half horse-power single-phase compensated series motor."

Clear, concise directions and careful drawings are features of this book. 79 Pages, 47 Illustrations.

Price, $1.00

BOOK DEPARTMENT

Popular Science Monthly

225 West 39th Street,

New York City

Fastening Binding Posts to Carbon Electrodes in Cells

IN wet cells having carbon electrodes it is necessary to have the binding-post substantially joined to the carbon, and the corroding action of the electrolyte, due to its creeping up and through the porous carbon on the binding-post where the latter joins the carbon, must be avoided.

To do this is quite difficult, but a good method is as follows: Drill a hole in the carbon the size of the binding-post screw, then enlarge the hole at the bottom, making it cone-shaped. This hole is then filled with a melted alloy consisting of two parts bismuth and one part tin, and the binding-post is screwed in before it becomes too hard. On cooling the metal expands, making a tight and substantial joint.

To prevent capillary attraction and creeping of the electrolyte up the carbon, the top of the binding-post in place should be dipped or soaked thoroughly with paraffin, wax, or insulating varnish.—PETER J. M. CLUTE.

A Portable Tool-Board for a Shop or Plant

IN a large refrigerating plant where repairs are likely to be made in any room or division at any time, the workmen found that it was almost impossible to select the proper tools necessary for each job and take them to it. As

The tools are hung on the sloping sides and the whole is mounted on casters

there was only one set of tools in the plant, they were arranged on an inverted V-board mounted on a base having truck casters.

When a breakdown occurred in any part of the plant, it was only necessary to push the tool-board to that spot, and all tools necessary were at hand.

Preserving the Luster on the Surface of Silverware

SILVERWARE may be kept bright and clean by coating it with a solution of collodion diluted in alcohol. The articles should be placed in hot water, after which they should be thoroughly dried before applying the coat of collodion. Care should be taken to have the coating very thin, so that when dry it is almost invisible.

"Let's Go!"

That has always been the call of men who ride Harley-Davidson motorcycles, the call of leaders. They never lag. In war or in peace, the rugged dependability of the Harley-Davidson has drawn toward it riders of the trail-blazing type.

During the war every Harley-Davidson motorcycle made became the mount of a man in khaki. These men and their Harley-Davidsons have established unequalled records for continuous performance; records in economy, upkeep, and dependability of service; records no after-the-war business or personal demand can afford to ignore.

"Ask the men in the service—they know"

Harley-Davidson Motor Co · Milwaukee · Wisconsin ···

Protecting the Automobile Top from Tearing

AN arrangement designed to prevent the tearing and stretching of automobile tops consists of a pair of coil springs placed between the top and the wid-screen. The jerks and strains

A compressed coil spring with buckle and snap to take the strain from the cloth

occasioned by driving over ruts and rough roads are taken up by these springs instead of being transmitted to the nails that fasten the top to the rear of the machine. All pulling out and tearing away of the top is thus avoided.—J. S. CHAPMAN.

Open the Automobile Bonnet for Hill-Climbing

WHEN driven in a mountainous country, especially in summertime, the Ford engine has a habit of boiling madly. This is, of course, due to the amount of low gear work it has to do: the slow rate of speed does not permit sufficient volume of air to pass through the tubes to cool them.

Taking off the bonnet is an expedient frequently resorted to by Ford drivers, but it is an awkward item to stow away. If left at home, the driver is apt to need it to protect the spark plugs and motor from splashes when fording a stream or when passing through mud-holes. It is not always possible to stop and cool down every few miles, and a good compromise may be effected by opening one side of the bonnet only (naturally the side from which the wind is blowing). The open side may be tied back securely with a piece of stout string, and rattling may be prevented by placing a piece of felt in the fold at the forward end.

Running Direct-Current Motors on Alternating Current

THE illustration shows a method which will be found useful in running D. C. motors on A. C. circuits. Disconnect the field winding from the armature A and connect the brushes of the machine together. Then arrange the field F of the motor in

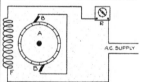

Wiring arrangement for running a direct-current motor on alternating circuits

series with a variable resistance R and the source of available alternating current. The resistance should be low for starting, but may be increased when the motor is running at full rated speed.

A Home-Made Candlestick that Will Not Upset

THE half of an old metal globe or a cup-shaped section of light metal can be turned into a convenient candlestick by denting in four spots on the sides as shown and then pour-

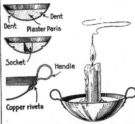

An old metal globe makes this attractive base for a candlestick

ing in enough plaster of Paris to form a candle socket of sufficient depth. The dents produce projections on the inner surface to hold the plaster. Before the plaster has set press a new candle into the center of it and let it remain there until the plaster is hard. This forms the socket. A more finished effect can be produced by riveting sheet metal handles on as shown in the illustration.—JAMES M. KANE.

A Switch for an Adjustable Condenser Circuit

THE device described and illustrated herewith is an easily constructed switch for use with an adjustable or variable condenser. Its construction and details are simple, and it is so built that it will not get out of order very easily.

The square base is made of wood, and the movable contactor consists of a large fiber washer and half of a brass washer, as shown, so fitted that there will be an even surface. The contacts, which are connected underneath

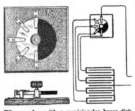

Fiber washer with a semicircular brass disk mounted on a wood base for a condenser circuit

to the condenser terminals, are made of brass strips cut and bent into shape, to bear on the movable washer. The method of connecting the switch to the condenser can readily be seen here.—PETER J. M. CLUTE.

Apparatus for Sending Wireless Messages from Airplanes

PICTURE yourself speeding aloft in a wireless-equipped airplane with two officers of the Signal Corps to obtain information that will silence an enemy battery. The day is clear, but it is difficult to locate the guns because of the camouflage. Sud-

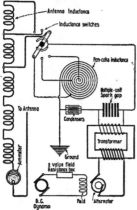

Diagram of connections of a Telefunken airplane wireless as used by the Germans

denly a bomb exposes a suspicious bit of landscape.

"There they are!" cries the observer. The others look in the direction in which he points. Far below them see what he has discovered—the well-masked muzzles of the guns that have caused so much destruction. Even the soldiers manning them are momentarily in sight.

The observer quickly turns to his wireless set and signals headquarters, telling just in what direction he has calculated the gunners' fire should be directed. Almost within the instant the hand of the operator leaves the transmitting key a veritable hail of shells begins to fall about the battery. They just fail to reach their mark, however, so the observer sends another message to headquarters—new directions for aiming. Then the fire is swung slightly to the left and directed squarely on the objective. The wireless has accomplished its mission, and the airplane descends to the earth.

An inspection of the radio apparatus in the airplane shows that it is somewhat light. Mounted on the forward part of the flying machine is a tapering metal case, hardly eleven inches long and about six inches in diameter, yet this contains an alternating-current generator of two-thirds horsepower and a direct-current dynamo to excite the alternator. When the airplane is in operation the propeller is driven

around 4,500 times a minute. The dynamos on the same shaft whirl around with it, and are thus made to go around ten times as fast as the wheels of an automobile traveling at forty miles an hour.

The alternator does away with an interrupter on a spark coil. The rapidly reversing current is led directly into the transformer primary. The current emerging from the secondary winding, now stepped up to a very high voltage, finds itself first charging the plates of the spark gap and then overwhelming the air resistance and breaking through in the form of an electric spark. The condenser in the circuit shunting the gap has in the meantime received a charge also, and the spark has helped to close the circuit containing the pancake inductance, the gap and the condenser.

Following the rest of the circuit, it is observed that the pancake is tapped directly onto the antenna and ground. The antenna consists of a reel of wire back of the observer's seat. This is unwound as the airplane takes the air and the antenna is lowered about 135 feet. A metal bobbin on the end of it keeps this wire from fluttering as it is swept almost horizontally in a long graceful curve by the wind. The "ground," on the other hand, is merely the wire stays between the planes and inside of the canvas covering of the wings.

The controls by which the operator tunes his set are on top of the instrument case mounted at the front of his compartment. With the first control he is able to change the power from the alternator from low to high as he gets farther and farther away from his lines. When the switch is thrown to high, the spark-gap length is increased automatically, so that he may be sure that the spark will work as efficiently as before. Throwing the next control simply adds or subtracts

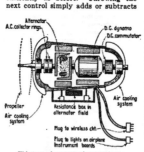

This type of generator has two dynamos mounted on the same shaft

some of the pancake's inductance so that any of the three wavelengths of 150, 200 or 250 meters may be quickly obtained. By this means, should the airman wish to report a maneuver he has discovered to a station, other than his own, he changes to the different

wavelengths of that station. Every time he does this he varies the inductance in the antenna circuit slightly until the ammeter in the same circuit shows that the greatest amount of power is being radiated.

This, then, is what you would have to be familiar with if you were an officer of the Signal Corps detailed to airplane observation work. But then, you may say, you have not been told how the receiving is accomplished. The answer to this is that it is not the usual practice. It can be done, but the added incumbrance to the aviator and the necessity for increasing the complexity of the system to overcome the noise from the engine and the rush of the air are advanced as arguments against it. The observer is given his orders before ascending. These he must fulfill to the letter; furthermore, he is expected to use initiative in the "tight places"—when, for instance, he is menaced by anti-aircraft guns or enemy wireless men "Jam" his messages.—LLOYD KUH.

Premonition Trick—Which Was the Last Coin Counted?

ARRANGE 19 coins in the form of the letter P. Tell your friends that while you are out of the room they are to think of any number and count it out on the coins. You will return when they are ready, and tell exactly what coin was last counted.

Of course, they must begin counting at X, and following the upright in the letter to Z, out toward V, then back down to Y, where the count goes out toward M.

The letter P is formed with the coins

Suppose the number should be 12. Begin at X, and count 12 coins, which takes the count to V. Now start back and count 12, which brings the count to M. Take any number you desire, and do the same, and you will find that the count always comes at M.

The location of M will always be easily found, because there are as many coins between Y and M as there are between Y and X. Of course, it is necessary to change the number occasionally between X and Y, so that M will not come in the same place every time.

Even if the number chosen is above 19, the trick will work, and the back count will come at M for the last coin. It makes a very puzzling trick for those who do not know how it works. Tell your audience that the coins must be arranged in the shape of the letter P because you tell which is the last coin by the instinct of premonition.—F. E. BRIMMER.

6,003 Burlingtons in the U. S. Navy—

6,003 Burlingtons have been sold to the men aboard the U. S. battleships. Practically every vessel in the U. S. Navy has many Burlington watches aboard. Some have over 100 Burlingtons. The victory of the Burlington among the men in the U. S. Navy is testimony to Burlington superiority.

A watch has to be made of sturdy stuff in order to "make good" on a man-of-war. The constant vibration, the extreme heat in the boiler rooms, the cold salt air and the change of climate from the Arctic to the Tropical are the most severe tests on a watch. If a watch will stand up and give active service aboard a man-of-war, it will stand up anywhere.

21-Jewel Burlington $2.50 A Month

And yet you may get a 21-jewel Burlington for only $2.50 a month. Truly it is the master watch. 21 ruby and sapphire jewels, adjusted to the second, temperature, isochronism and positions. Fitted at the factory in a gold strata case, warranted for 25 years. All the newest cases are yours to choose from. You pay only the rock-bottom-direct-price—positively the exact price that the wholesale dealer would have to pay.

See It First!
You don't pay a cent to anybody until you see the watch. We ship the watch to you on approval. You are the sole judge. No obligation to buy merely because you get the watch on approval.

Write for Booklet!

Put your name and address in the coupon or on a letter or post card now and get your Burlington Watch book free and prepaid. You will know a lot more about watch buying when you read it. Too, you will see handsome illustrations in full color of all the newest cases from which you have to choose. The booklet is free. Merely send your name and address on the coupon.

Burlington Watch Company,
19th Street and Marshall Blvd., Dept. 1201 Chicago, Illinois
Canadian Office: 355 Portage Ave., Winnipeg, Man.

Burlington Watch Co., Dept. 1201,
19th Street & Marshall Blvd., Chicago, Ill.

Please send me (without obligations and prepaid) your free book on watches with full explanation of your cash or $2.50 a month offer on the Burlington Watch.

Name...

Address...

Electrical Devices and How They Work

XIII.—How the resistance encountered in the conductor generates heat

By Peter J. M. Clute, B. C.

WHEN a current of electricity passes through a conductor, it encounters a certain amount of resistance. In overcoming this resistance heat is generated. Along with this production of heat there is also frequently light or motion. The heat developed is proportional to the square of the current, directly proportional to the resistance and directly in proportion to the time. Thus, if the current is doubled there will be four times as much heat developed as was previously generated; and, if the current remains the same and the resistance is doubled, there will be twice as much heat as before.

When electricity flows through a conductor the work in joules is equal to I^2Rt, where I represents amperes, R ohms, and t seconds. All this work is converted into heat energy, which raises the temperature on the conductor and its surroundings.

Figure 1

A simple form of an electric furnace consists of a crucible with a fire brick housing

Transmitting the Current

In the generation and transmission of electrical energy this production of heat is very undesirable, and is avoided as much as possible by using conductors of low resistance or by transmitting the energy at high pressure and correspondingly low current. Ordinarily the transmission work, the size of the conductors to be used, is determined by the allowable pressure drop rather than by the heating effect of the electric current; but it is sometimes necessary to consider also this heating effect.

The fusing effect of a current depends on the readiness with which heat can escape from the wire. If a very short wire is clamped between terminals, heat will escape to the wire; if a fuse is installed where air circulates freely, the air currents will carry away the heat. For these reasons, fuses must be of sufficient length so that the heat imparted to the terminals cannot appreciably change the melting-point; they must also be installed where air currents cannot affect them. Fuses, therefore, are usually 1 in. or more long, and are inclosed.

Difficulty of Good Insulation

In the absence of air, a conductor will carry a much larger current without fusing than if air is present. As a

consequence, in rheostats and in electric-heating apparatus, where a high-current density or an intense heat is desirable, the wire is embedded in cement, enamel, or other substance, which not only insulates the conductors, but also excludes the air from around them. The incandescent lamp affords an example of the advantage of excluding air from a highly heated conductor. If even a very small amount of air remains in the bulb the life of the lamp will be much shortened; if a filament should be exposed to the air it would immediately be consumed.

The resistance wire in rheostats and electric-heating devices, if properly protected from air, may be operated at red heat without material injury; but this is seldom done, because it is difficult to maintain good insulation at such high temperatures, and, any way, such intense heat in these appliances is seldom necessary.

In every kind of work that electrical energy is called upon to perform, resistance, and consequent waste by heating, must always be reckoned with; and in every kind of work, except electric heating, it must be reckoned as a loss. Electric heat is the one kind of work that electricity does in which resistance is a good thing—indeed, the essential thing.

Uses for Electric Heat

In the electric light—arc, incandescent, or any other kind—it will be noted that the light comes from solid or gaseous conductors that are raised to a white heat by passing a current through them. In order to get the light from such electric lamps, a good deal of heat energy is generated along with it—in fact, most of the electrical energy is transformed into heat and only a little into light. In making electric heating apparatus, however, it is not light, but all heat, that is desirable; and the effect of electrical friction, which is such a waste and loss in all other electrical devices, is here the very essential thing. Thus, as far as efficiency is concerned, electric heating devices are superior to any other kind · of electrical apparatus.

Electric heat is now employed for the following principal uses: (1) to operate electric furnaces in the art of electro-metallurgy; (2) in electric welding; (3)

in the electrolytic forge, or tempering bath; (4) for air-heating devices; (5) for water-heating devices; (6) heating appliances for domestic use; and for numerous other miscellaneous appliances.

A great variety of electro-metallurgical processes can be accomplished by the use of the electric furnace. The electric furnace has the advantage of producing a very high temperature and of being under easy control. Since the carbon serves merely as the reducing agent, and not at the same time as fuel, its quantity can be regulated to meet the requirements of whatever chemical reaction it is desired to secure.

When the current is made to flow across an air gap between two carbon electrodes, an electric arc is produced. The temperature of the arc is the highest obtainable, being in the neighborhood of (3500° C.); and in the electric furnace, in which the arc is confined in an inclosed space, any known substance can be melted or vaporized. Carbon is nearly always used for the electrodes, as it will best withstand the heat.

A Simple Electric Furnace

A simple form of an electric furnace is shown in Fig. 1. It consists of a crucible of refractory material surrounded

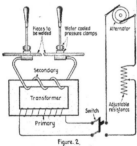

Figure. 2

A diagrammatic view of the Thompson process, the one most commonly used for welding

by fire-brick, and is covered with a fire-clay slab. Carbon rods enter from each side and form the electrodes, the arc being started by sliding one carbon in until it touches the other and then withdrawing it.

The great usefulness of the electric furnace is due to the ease with which its temperature can be regulated, also to the intense heat obtainable. There are three general types: the reflecting arc furnace, in which the heat is reflected upon the reflecting surface; the direct heating arc furnace, in which the materials treated surround the arc that imparts heat directly to the charge; and the incandescent furnace, in which either the charge or the furnace wall forms a part of the circuit, and by the resistance offered to the passage of the current produces the heat effects required. The direct heat-

ing and the incandescent furnace are the types most frequently used.

In the ordinary process of welding materials the parts to be united—say two bars of iron—are heated in the forge and then hammered together. Electric welding works much more quickly, and the joint is as neat. Moreover, it can weld certain metals that cannot be welded by the old method, and it can weld together two different metals.

The Welding Process

The process consists in simply pressing together the ends of the pieces to be welded and passing a very strong current through the point of contact. An alternating current is generally used, because a heavy current at low voltage is more easily secured by using an alternating current transformer of special design rather than by employing a direct-current dynamo. The heavy current finds a resistance to its flow in passing from one bar to the other where the bars touch, and generates an intense heat at that point. The metal becomes white hot

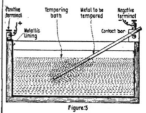

Figure 3
A cross-section of an electrolytic forge, showing the electrode in the water bath

in an instant, and it is only necessary to force them together to make a true welded joint. A special machine, called an electric welder, is used. It has heavy copper clamps to grip the pieces to be welded, and an arrangement of levers to press them together at the proper moment. The Thompson process, which is more generally used than any other, is illustrated in Fig. 2.

Only a very low voltage is needed in the secondary circuit, but a current as high as 60,000 amperes to the square inch may be necessary in welding some metals, as, for example, two copper pieces. A low frequency—50 cycles or even less—is preferred, especially for heavy work where the current density is very great.

The time required for making a weld varies inversely with the amount of power supplied; that is, the greater the power the shorter the time of the weld, and *vice versa*. Metals that are deteriorated by being heated, such as brass, copper, and tool steel, must be welded rapidly. The pressure must be great enough to crowd out from the weld all metal harmed by the heat

Play Any Instrument a Week on Free Trial!

YOU may take your choice of any of the instruments in our big, new catalog and we will send it to you for a week's free trial. We want you to compare it with other instruments—and to put it to any test. We want you to use it just as if it were your own. Then, if you wish, you may return it at our expense. No charge is made for playing the instrument a week on trial.

WURLITZER

200 YEARS OF INSTRUMENT MAKING

Convenient Monthly Payments If you decide to buy—you may pay the low rock-bottom price in small installments, if you wish. $4.00 a month will buy a splendid triple silver-plated cornet. You will find over 2,000 instruments in our catalog. Every one is backed by our guarantee. Every one is offered to you on the same liberal plan. The name of Wurlitzer has been stamped on the finest musical instruments for 200 years. Wurlitzer has supplied the United States Government with trumpets for 55 years. *We are specially prepared to assist in the formation of bands or orchestras.*

Send the Coupon

Send your name and address on the coupon (or in a letter or post card) and get our new catalog. It takes 160 pages to show you the instruments from which you may choose. The catalog is sent free, and without obligation to you. Merely state what instrument interests you—and send your name. Don't delay—do it today, RIGHT NOW.

The Rudolph Wurlitzer Co.
So. Wabash Avenue, Chicago—Dept. 1201—E. 4th St., Cincinnati, O.

The Rudolph Wurlitzer Co.
Dept. 1201
E. 4th Street, Cincinnati, Ohio
S. Wabash Avenue, Chicago, Ill.

Gentlemen:—Please send me your 160-page catalog, absolutely free. Also tell about your special offer direct from manufacturer.

Name_____

Street and No. _____

City_____State_____

I am interested in_____
Name of instrument or band or orchestra

An electrolytic forge, or tempering bath, consists of a vessel lined with metal containing water or a suitable solution. The solution is made the positive electrode of a direct-current generator, while a piece of metal to be heated and tempered is made the negative electrode. This device is illustrated in Fig. 3. The liquid is contained in a vessel that has a metal lining to which the positive side of the circuit is connected. The piece of metal to be heated rests on a contact-bar, to which the negative side of the circuit is connected and extends to the liquid.

The simple action of the tempering bath is as follows: When the metal to be heated and tempered is plunged into the liquid and adjusted to the rod, a current begins to flow through the liquid to the rod, and a layer of hydrogen gas immediately forms around the submerged portion. This layer of gas introduces so much resistance between the liquid and the metal that intense heat is thereby developed at the surface of the metal. By adjusting the current strength and the time the current is allowed to flow, any required degree of heat can be obtained, even to melting the metal.

Figure 4

Non-inductive effects are produced by making the current flow in a zigzag path

Suitable insulating shields placed over portions of the metal prevent the development of heat on surfaces that are not to be tempered. The heating is under such perfect control that the tempering may be carried to any desired depth from the surface of the metal.

Electric Heating Systems

General heating by electricity can be carried out by three different systems: by radiation, in the way that the sun warms the earth; by convection, by which is understood the direct heating of air by contact with a heated surface; and by conduction, which is the effect produced upon solid objects that are in contact with others at a higher temperature, the heat traveling along the material and warming it. The last named system is made use of only to a small extent, but a certain amount of conducted heat is given off from any solid body that touches another solid body at a lower temperature, and every heating appliance gives out a greater or less proportion of its energy in the form of conducted heat.

In all electric heaters the heat is developed by the material contained in the appliance passing through some form of resistance. They may be broadly classified as follows: first, luminous heaters, or radiators, second, non-luminous heaters, or convectors; third, red glow heaters.

In the luminous type of heaters the heat is obtained from two or more long cylindrical carbon filament lamps, especially designed for the production of heat rather than light. These lamps are mounted in a metal case opened at the front with a polished metal reflector in the back. Heat is imparted to the containing bulbs of the lamps, and thence by radiation to the surrounding air.

How Heat Is Obtained

In the non-luminous type, heat is obtained by the passage of current through bare wire or metal strips of high resistance, wound on tubes or frames of porcelain or other highly refractory substance, each tube or frame forming a heating element. Several heating elements are contained in a metal case from which they are insulated, and they are coupled together by short wires so as to form continuous lengths of high-resistance material between the two terminals of the heater.

In the red glow heaters the heating element is formed of a thin high-resistance wire spiral contained in a quartz tube. Several of these glowers are fixed in a frame, and the frame is mounted in the open front of the metal containing case. Quartz is as transparent to heat as to light, hence the bright red glow of the spiral wire inside the tube is not only visible, but the heat from it passes out to the surrounding air. A large surface of red-hot quartz is thus obtained, giving an efficient radiator of cheerful appearance.

All electric heating devices may be classified into lighting circuit and heating circuit devices. The lighting circuit devices are those that take about 5000 watts or less and may be connected to the ordinary branch circuits without any special wiring. The heating circuit devices require special circuits, as ordinary branch lighting circuits are not of sufficient capacity. In such devices the heating circuits are arranged as closely as possible to the surfaces to be heated, so as to make the efficiency of conversion from electricity into useful heat as high as possible. Generally, each manufacturing company has adopted a distinctive method of making and insulating the resistances.

All heating resistances for use with alternating current should be non-inductive, as the production of heat depends only on the square of the current and the ohmic resistance; inductance would cause voltage losses that would result in no additional heat. Non-inductive effects are produced by making the current follow a zigzag path, as suggested in Fig. 4; or, if the resistance is in the form of a helix, by making the winding such that the current must travel an equal number of times each way around the helix.

The applications of electric heat are very numerous. It is important to remember that the "same kind of electricity" is used alike for lighting and

for heating, although different prices may be charged for the two services. The reason for a distinction in prices is based on economic considerations too complicated for discussion, the general idea being that, as a rule, when the demand for artificial light is least it pays the central stations to offer energy at low rates in order to encourage its use at hours when little power would otherwise be demanded.

(To be continued.)

An Automobile Horn Attached to the Frame Out of the Way

TO place the horn out of the way of the driver, an extension-rod for the push-knob was applied as shown in the illustration. The horn is placed

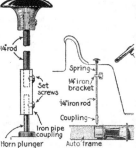

¼ rod — Spring — ¼ iron — Set screws — ¼ iron rod — Coupling — Iron pipe coupling — Horn plunger — Auto frame — bracket

The automobile horn is placed on the frame out of the way and is worked by an extension rod on the push-knob

on the frame out of the way, and the horn plunger is lengthened with a rod which carries the push-knob at the top. A bracket is attached to the seat arm and a coil-spring placed between the bracket and the knob.—P. P. AVERY.

How to Cut a House Number in a Board

THE house number shown is for the consideration of those who like something a little different. It is made of a piece of board 5 in. wide, 2 ft. long, and ¾ in. thick. The numbers are sawed out with a compass or key-

The house number is cut from a board like a stencil and hung in the porch

hole saw. The board is for hanging from the front edge of the porch; in such a position the number may be read as easily at night as in the day. Care should be taken to have the ties that hold the central portions of the figures run with the grain of the wood, as shown. If put in across the grain, they are, of course, easily broken.—EDWARD H. CRUSSELL.

A Baseball Batting-Machine

The baseball fan may still exercise himself with the bat when winter chills compel him to stay inside

By A. D. Goodrich

AT a summer camp of an athletic club there was no suitable field for playing baseball, and some of the boys manufactured a batting-machine. With its help they could play their ball games on a small lot back of the camp or in the gymnasium on rainy days. Any number could play; they seemed to get about as much sport out of it as in playing the regulation game, and the machine was in almost constant use. Such a machine can be used in the winter for indoor exercise.

The materials for this batting-machine are easily obtainable at a reasonable cost, and any boy with a little knack in using the most common carpenter tools can make one in a few hours. The upright or score-board of the machine is made of good clear pine planed on one side, 12 or 14 ft. long, 4 in. wide, and ⅞ in. thick, supported by a base with braces, as shown in Fig. 1.

The base framework should be about 3 ft. square, with the bottom boarded over to strengthen it, and stones or bags of sand laid on it to weight it down and steady it. At the top a 6-in. strip is nailed to the back of the upright, as shown at A, Fig. 4. This strengthens the projecting piece B, to which is fastened a small pulley, C, that carries the line D. The line or cable should be strong, fairly heavy, but easy running cord—a good trolling fish-line is about right.

The weight at the lower end may be either an iron bolt or a piece of lead cut or molded to a pear-shape, and, when fastened to a line that is fully unwound, hangs about 2 ft. from the ground.

He is striking the ball to make it record hits on the marked upright

The bell may be one from an old alarm-clock or the top of an old bicycle bell. This is fastened to the face of the score-board E near the top, the stem coming at one side of the line, which must pass freely under the bell.

First, the bracket extending from the standard and supporting the striking arm is made. The horizontal boards are about 4 in. wide and 22 in. long, fastened 42 in. from the ground to the standard with a reinforcing piece of board at the back and a similar piece connecting their forward ends. Two braces run diagonally up to the standard and are nailed to it.

The striking arm is a piece of wood 18 in. long, 3½ in. wide, and 1¼ in. thick. Beginning a short way from either center, its ends taper so that the last 2 in. of their length may be whittled round to go into the holes bored in two croquet balls. The balls may be held in a vise for boring the holes, and the holes should be a drive fit for the tapered stick. Glue should be applied before driving them in place. One of these balls serves as a striking ball, and is shown at G, Fig. 3. The ball G is padded by placing strips of cloth or cotton batting around it and winding the whole with twine.

The striking arm shaft J is merely a 2-in. square of hard wood long enough to reach through the bracket boards

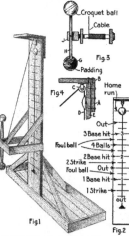

 Labels: Croquet ball, Cable, Fig.3, Padding, Home run), Out, 3 Base hit, 4 Balls, 2 Base hit, 2 Strike, Foul ball, Out, 1 Base hit, 1 Strike, out, Fig.4, Fig.2, Fig.1

Details of the standard and ball for striking and base for holding it firm to the floor

and afford a projection at either end, as shown. This projection should be of sufficient length at the striking arm end to permit the balls to rotate without striking the bracket board. At the other end it is long enough to place an iron washer over it and to drive a wire nail through the shaft to hold it in its bearings. The head of this nail should be removed.

Of course, that part of the shaft passing through the bracket boards and within them is made round, and the holes in the boards are bored large enough to permit the shaft to rotate freely but not loosely. These bearings should be kept well greased. The shaft engages the striking arm by a square hold cut through the center. A nail driven into the shaft from the edge of the striking arm secures these parts.

The cable end is tacked to the shaft at one side, and so wound that it will wind up when the striking ball is hit. This ball, if it does not adjust itself, must always be turned straight downward before it is batted again.

Any number of players, equally divided to represent sides, can play the game. A toss is made between the two captains to see who will have the choice of ins or outs; the outs stand aside until three men of the ins have retired themselves, as in the regular game; then the outs take their turn at the bat.

It is well to have the bases marked out near the machine and the players take their places as the score is shown on the board at each hit: that is, if the man at the bat hits so that the marker goes to 2-base hit (Fig. 2) he at once takes his position on second base; if the next man makes a 2-base hit he takes his place on second base, and the first man steps home and scores a run; and so on through the game. In other words, the game is played exactly the same as the regulation game, except that the machine is used to take the place of the pitcher and catcher, and as soon as the boys get used to hitting and scoring with the machine plenty of sport and also beneficial exercise will be the result.—A. D. GOODRICH.

An Eye-Bolt Made of a Piece of Sheet Metal

AN EYE-BOLT being needed in the construction of a small boat far from any place where one could be purchased, the workman made one from a piece of metal, as shown.

A strip of sheet metal was bent double in the exact center, and the ends again bent at right angles to the double shank left by the first bending. A hole of sufficient size was then drilled in the ends for screws.—L. B. ROBINS.

The metal bent to substitute an eye-bolt

 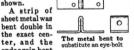

QUICK-ACTION ADVERTISING

HERE READERS AND ADVERTISERS MEET TO TRANSACT BUSINESS

Rate 20 Cents a Word, payable in advance. 10% Discount for Six Insertions. February issue closes December 20th

AUTOMOBILES AND ACCESSORIES

HYDRONIZER—Insures clean plugs, consumes carbon, saves gasoline, saves repairs and is easy to sell. Money back guarantee. Fits all cars. Fords do 34 miles to gallon. Thousands in use. Have attractive proposition for energetic and responsible agents, also unique advertising. Agents' price permits wholesaling and retailing. Write for free literature. Friestedt Manufacturing Company, 2933 W. Lake St., Chicago, Illinois.

CONVERT your bicycle into a motor-cycle at small cost by using Shaw Motor Attachment. Catalog free. Shaw Mfg. Co., Dept. 180, Galesburg, Kansas.

PATENTS for automobiles are in great demand. We have special Automobile Department in charge of experts who are in touch with demands of trade. Write for free Illustrated Guide Book, "How to Obtain A Patent." Victor J. Evans & Co., 172 Ninth, Washington, District of Columbia.

BATTERY Charging pays big profits. City currents or gas engine operates. Easy terms. Hoberts, Troy, Ohio.

INSYDE Tyres, inner armour for Automobile Tires, prevent punctures and double mileage of any tire. Liberal profits. Details free. American Accessories Co., Dept. 97-A, Cincinnati, Ohio.

TIRES—Double Tread Guarantee for good service. Absolutely Puncture-Proof. Big—Strong—Extra heavy 30x3 tire, $5.50; 30x3½, $7.00; 33x4, $10.00; 34x4, $10.75; 36x4½, $13.00; 37x5, $14.00. Big saving on other sizes and tubes also. Trade in your old tires. Discount to dealers. 10% deposit required on C. O. D. orders. Send for list now! State size and bead of tire. Max Liben & Co., 205P West 48th Street, New York.

TIRES At Wholesale—Send For Prices on guaranteed tires and tubes. Lowest prices on earth. Overton Tire Co., Box P, Oelwin, Iowa.

"GAS-SOL" gives you 15 to 40% more mileage and eliminates carbon. Money back Guarantee. 100 tablets treat one hundred gallons gasoline. Price One Dollar. Live agents wanted. Gas-Sol, Dept. B, Wakefield, Mass.

MORE Power, less fuel, no carbon. No mystery, plain facts, results guaranteed. Write for booklet. No-Leak-O Piston Ring Company, Baltimore, Maryland.

VULCANIZING auto-tires is now a growing and profitable business. Easy to learn. Instruction book $1. Plants $50 to $300. Details free. Equipment Company, 302 8th Street, Cincinnati, Ohio.

TIRES ¼ less. Buy direct at manufacturer's price. Guaranteed 6000 miles. Shipped prepaid on approval. Motorist Agents wanted. Write today. Give size of tires. Liberty Tube & Tire Co., 842 Grand Avenue, Kansas City, Missouri.

AUTO Motors and Supplies. Buick—Hupp—Franklin—Michigan—Everett—Both water and air cooled Motors $30.00 each and up. Bosch Magnetos—$10.00 each and up—Presto Tanks $5.00—Coils $3.00—Carburetors $5.00—Electric Head Lamps $4.40—Spot Lights $3.40—Motor Horns $2.85. Air Compressors, etc. Write for Bargain List second hand auto accessories. Johnston, West End, Pittsburgh, Pennsylvania.

WILL purchase good automobile accessory, or handle same on royalty basis. F. R. Company, Box 67, Station B. Cleveland, Ohio.

MR. ADVERTISER: If you knew as much about the pulling-power of Popular Science Monthly as the advertisers represented in its columns, your advertisement would be in this space. We will be glad to give you some interesting information about this department. May we? D. W. Coutlee, 225 West 39th Street, New York City.

FORD ACCESSORIES

FORDS Start Easy in Cold Weather with our new 1919 carburetors. 34 miles per gallon. Use cheapest gasoline or half kerosene. Increased power. Styles for any motor. Very slow on high. Attach it yourself. Big profits to agents. Money back guarantee. 30 days' trial. Air-Friction Carburetor Co., 500 Madison, Dayton, Ohio.

DEALERS: If it's for a "Ford" car we have it. Write for price list. Universal Motor Supply Company, White Plains, New York.

DEMOUNTABLE Outfits complete, $12.50. Demountable sets, $5.50. Good agents' proposition. Kable, West End, Pittsburgh.

SUNLIGHT for Fords! Bright light all speeds, with dimmer, $2.50 guaranteed. 100% Profit for agents. Sunlight System Company, 1508 Lytton Building, Chicago, Illinois.

SPECIAL prices on Ford Tires: 30x3, $10.50, tube, $2.25; 30x3½, $12.50—tube, $2.50. John A. Martin, Aberdeen, South Dakota.

ELECTRICAL

MAKE Dry Batteries. Simple, practical instructions with blue print, 25 cents. Dirigo Sales Company, Bath, Maine.

INVENTORS. Send sketch and description of your invention for advice regarding patent protection. No charge for this service. 20 years experience. Prompt personal service. Patents we secure advertised without charge in Popular Mechanics Magazine. Particulars free. Talbert & Talbert, Patent Lawyers, 4675 Talbert Building, Washington, D. C.

GENERATORS, motors, electrical supplies. Illustrated list free. Hyre Electric Company, 629 BK South Dearborn Street, Chicago.

WELDING

WELDING Plants, $25.00 to $300.00. Designed for all purposes. Small cash payment, balance three to six months. Every mechanic or shop should have one. Bermo Welding Co., Omaha, U. S. A.

MISCELLANEOUS

ELECTRICAL Tattooing Machines, $3, $5, and $7. Catalogue for stamp. J. H. Temke, 517 Central Pam., Cincinnati, Ohio.

AVIATION

PATENTS wanted in Aeronautics. U. S. Government spending millions in this class for Army and Navy. Our Aeronautical Department will give full information on this subject. Write for our free book, "How to Obtain A Patent." Victor J. Evans & Co., 179 Ninth, Washington, District of Columbia.

MEN: Learn aviation, motor mechanics and aeroplane building. Earn $1.00 to $2.00 hour when skilled. Wonderful future. Big government plans. Write. Moler Aviation School, 105 South Wells, Chicago, Illinois.

BUILD that new airplane now for the coming flying season. Write for catalogue "K" and bargain motor. Send 12c in stamps to America's oldest aeronautical supply house. Heath Airplane Company, Chicago.

BUILD the Marinette Monoplane; use motorcycle engine. Stamp for particulars. M. Meyler, 218 East Main Street, Kalamazoo, Michigan.

TRADE SCHOOLS

HILES' Watchmaking and Engraving School, the largest and best equipped school in the West. 717 Market Street, San Francisco, California.

Popular Science Monthly,
225 West 39th Street,
New York, N. Y.

Gentlemen:

Your magazine is one of four that have netted us almost 2,000 inquiries for our agency proposition during the past few months. We receive orders from a large percentage of those with whom your magazine has placed us in touch.

In addition to our agent advertisement, we have been using one under the heading of "Adding Machines" to secure orders from general trade. Most of the people that are interested in this advertisement send their dollars along; and of those that merely inquire, our follow-up succeeds in selling 90%. Our ads in your magazine are so satisfactory, that we have decided to use them for everything to which they can be applied.

Very respectfully yours,

L. J. LEISHMAN COMPANY.

DUPLICATING DEVICES

"MODERN" Duplicator—a Business Getter. $1 up. 50 to 75 copies from pen, pencil, typewriter; no glue or gelatine. 35,000 firms use it. 30 days' Trial. You need one. Booklet Free. J. V. Durkin & Reeves Company, Pittsburgh, Pa.

MR. ADVERTISER: If you knew as much about the pulling-power of Popular Science Monthly as the advertisers represented in its columns, your advertisement would be in this space. We will be glad to give you some interesting information about this department. May we? D. W. Coutlee, 225 West 39th Street, New York City.

MODELS AND MODEL SUPPLIES

TO avoid mistakes, inventors should have correct models made before filing for patent. Write Adam Fisher Mfg. Co., 183E, St. Louis, Missouri.

MANUFACTURING

SPECIALTIES of all kinds manufactured for market. Write Adam Fisher Mfg. Co., 183F, St. Louis, Missouri.

SOAP Saver! New invention! Patent applied for. For sale or royalty. Manufacturers write at once! N. I. Ward, 7 Myrtle Street, Bellows Falls, Vermont.

HAVE you an idea to develop, a machine to perfect, a product to be manufactured? Do you need special machinery to reduce the cost of your product or a first class model built, tools, dies or any kind of machine work? If so, send us a card for our illustrated booklet for manufacturers and inventors. Central Machine Works, 1911 North 12th Street, St. Louis, Missouri.

FORMULAS & TRADE SECRETS

FORMULAS—All kinds—Valuable Catalog Free. "Bestovall," Box 543-PS., Chicago.

WILL furnish a typewritten, guaranteed Formula or Trade Secret for any purpose for only 25c. State exact requirements enclosing coin. Raymond, 1354S Central, Cleveland, Ohio.

MOTORCYCLES, BICYCLES, SUPPLIES

MOTORCYCLES all makes, $25.00 up. New bicycles at big reduction. Second hand, $8.00 up. Motors, motor attachments, Cycle motors, Smith motor wheels, etc., $20.00 up. New parts to fit all makes carried in stock. Second hand parts good as new 50% discount. Expert repairing, on magnetos, generators, transmissions. Motors overhauled $10.00 up. Henderson motors our specialty. Write for big bargain bulletin. American Motor Cycle Company. Dept. 3, Chicago.

BARGAINS in used Harley, Excelsior, Henderson and Indian Motorcycles; condition guaranteed, $40.00 to $250.00. Write for terms and new bulletin. Charles A. Merkle, 162 North Clinton Avenue, Rochester, New York.

WANT to sell that motorcycle of yours? I have a method that has sold many. Write today for details; a postcard will do. D. W. Coutlee, 225 West 39th Street, New York City.

CONVERT your bicycle into a motorcycle at small cost by using Shaw Motor Attachment. Catalog free. Shaw Mfg. Co., Dept. 68, Galesburg, Kansas.

TRANSFER nameplates for motorcycles, etc. Samples and quotations to dealers only. Globe Decalcomanie Company, Jersey City, New Jersey.

GRANDFATHER'S CLOCKS

GRANDFATHER'S Clock Works $5.00. Build your own cases from our free instructions. Everybody wants a hall clock. You can make good profit building artistic clocks for your friends. We replace worn-out works in old clocks with works having chimes at money saving prices. Write for folder describing the most beautiful hall clock ever sold at $25.00. Clock Co., Nicetown, Pennsylvania.

MOTORS, ENGINES, MACHINERY

BLUE Prints, Castings, Small Engines, Steam, Gas, Circulars, Stamp. Universal Gas Motor Company, Monadnock Block, Chicago.

SMALL Motors and Generators: 1000 New Motors and Generators from bankruptcy stock—¼ H. A.C. $18.50 each—½ H.P. $30.00. Battery Charging Sets—Robbins & Myers New outfits all sizes, $30.00 each and up. Charging Lighting & Moving Picture Arc Generators $10.00 each and up. Motors for all phases of current. Immediate Delivery. Less than ½ regular prices. Write for late Bulletin. Johnston, West End, Pittsburgh, Pennsylvania.

GENERATORS, electric motors. All kinds. Free illustrated list. Hyre Electric Company, 629 GW South Dearborn Street, Chicago.

TELEGRAPHY

TELEGRAPHY and railway accounting taught thoroughly and quickly. Unprecedented demand for both sexes at big salaries. Oldest and largest school—established 45 years. Catalog free. Dodges Institute, Second Street, Valparaiso, Indiana.

TYPEWRITERS AND SUPPLIES

LARGEST stock of Typewriters in America. Underwoods, way under manufacturer's prices. Rented anywhere, applying rent on purchase price; free trial. Instalment payment desired. Write for catalogue 11. Typewriter Emporium (Estab. 1892). 34-36 West Lake Street, Chicago, Illinois.

NEW, rebuilt and slighty used Typewriters $8 up. Portable Machines $10 up. Write for our Catalog 25P. Beran Typewriter Co., 58 W. Washington St., Chicago.

ADDING MACHINES

WONDERFUL Adding Machine seven columns capacity, only one dollar. Adds and multiplies as fast as the fingers will move. Thousands being sold through demonstration. L. J. Leishman Company Dep't L., Ogden, Utah.

INVESTIGATE Marvelous Calculator Adding Machines. Substracts, multiplies, divides automatically. Work equals $300 machine. Price $10. Five year guarantee. Illustrated catalog and trial offer free. Calculator Corporation, Grand Rapids, Michigan.

ADVERTISING

YOUR advertisement in the classified columns of the Electrical Experimenter, Science and Invention, will reach the very class of men you seek. Circulation 100,000 net. Rate 6c word. The Electrical Experimenter brings positive results. For proof, address Classified Department, 233 Fulton Street, New York.

WHAT'S wrong with your sales letters and follow-up literature? Send complete set for criticism and revision. We'll make it right! Criterion Letter Service, West New York, New Jersey.

SPECIAL! Inch display advertisement 100 magazines, thrice, $8.00. Coast to Coast Syndicate, Atlantic City.

LABORATORY EQUIPMENT

EXPERIMENTERS. Equip a chemical laboratory. Investigate our complete lines of apparatus, chemicals, glassware, minerals and other laboratory supplies. Clarence Appel, Mathews Avenue, Knoxville, Pittsburgh, Pennsylvania.

FLAGS

FLAGS: 3x5 feet fast color American flag, $1.00. Also larger sizes. Allied and "Welcome Home" flags, all sizes. Jensen, 342 W. 42., New York City.

"TWENTIETH CENTURY BOOK OF RECIPES, FORMULAS AND PROCESSES"—This book gives you the information that represents the collection of years and the expenditure of thousands of dollars. Contains complete directions of making everything from simple glues and adhesive to the tanning of leather, the making of photographic materials, perfumes, soaps, plating of silver and gold, working of metals, etc. 800 Pages, 10,000 Selected Subjects. Price $3.00. Book Department, Popular Science Monthly, 225 West Thirty-ninth Street, New York.

Why You Should Use Both Eyes When Shooting

THERE is a great deal of contention among shooters as to whether it is better to shoot with one eye closed or both eyes open. The diagram will prove that it is best to shoot with both eyes open, since, in using one eye, there is somewhere in the range of vision a blind spot.

Hold the diagram at arm's length, close one eye, look steadily at the cross, then slowly draw the page toward your face. A position can be easily found where the bird in the picture will vanish. Open the closed eye and the bird will be clearly seen as before.

Everybody has a blind spot, and that may be where the aim should be

This shows not only that there is a blind spot when one eye is used, but that the blind spot for both does not coincide. Therefore both eyes should be used in shooting, so that the whole range of vision will be clear and distinct.

Suppose that you are about to shoot at a bird thirty yards away, and shut one eye; you are blind covering a space of several feet. If the bird flies through that blind spot he will fool you every time. Clay bird breakers and game hunters are learning to shoot with both eyes open to get the maximum efficiency for accurate shooting. Because a moving target has to be led with the gun muzzle, it is a mighty easy thing for the wing shooter to lose his bird because the lead placed his target in his blind spot.—F. E. BRIMMER.

Grinding Broken Hacksaw Teeth to Prevent Slipping

HACKSAW blades that have a few teeth broken out may be restored to usefulness without stripping the remaining teeth, by grinding the top off the adjacent on each side of the broken section, and grinding three or four teeth on a slant extending from the bottom of the broken ones to the top of the ones that are whole.

In grinding saws for special purposes, blades without any set are sometimes required. These may be made from stock blades by resting them flat on the emery-wheel and grinding the teeth down to the level of the body. Many workmen put the blade in the frame for such work, but this increases the breakage. The blade should be taken out of the frame and held in the hand, so that it is more or less self-adjusting. Strained in a frame as for sawing, the tension is such that the jar of the emery grains striking it breaks the blade just as if a hammer had struck it.

A Screw-Driver with Two Angle Blades

ONE of the most handy screw-drivers to have in one's tool-box is the angle blade, which has three edges. To make it, draw out the point

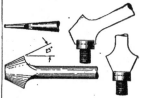

The three edges of this screw-driver blade make it adaptable for any kind of work

quite wide, and grind the edges at an angle of 25 deg. to the center line of the screw-driver.

When the side edges are used, the length of the screw-driver makes a leverage for starting or turning tight or close fitting screws.—EDWIN M. DAVIS.

Weighing Large Articles in Instalments

SOMETIMES one desires to weigh an article too heavy for the scale at hand, or too long to rest on the platform. In such cases, it is possible to resort to the trick of weighing first one end of the article and then the other. In the illustration the entire weight of the shaft C is supported at the two points A and B, and if there were two scales, one under each point, the combined amounts would equal the total weight of the shaft. With one scale the same result can be obtained by weighing first one and then the other and adding together the two amounts. To obtain the greatest accuracy, it is necessary that the shaft rest on the two A-shaped supports. The one not on the scale should be blocked up on a level with the scale platform D. A chalk mark should be made on each

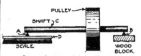

Supporting an article so that it can be weighed one end at a time

end of the shaft where it takes a bearing. One end should be weighed, and the amount recorded. The shaft should then be turned end for end, and placed so that the chalk marks correspond with the bearings. The second weighing should be added to the first, and the total will be the correct weight of the article.

A freight-car may be weighed on a scale that is too short or too light by weighing first one set of trucks and then the other.—W. H. SARGENT.

Gutter-Spout Directing Water into Two Trenches

FARMERS in irrigated districts frequently experience difficulty in making water flow into two different lateral trenches when the supply is obtained from one hydrant. Usually the soil under the spigot is washed away, and the water, seeking the lowest level, is soon flowing into one trench only.

A Western ranchman has effectively solved this problem, and at the same time has found serviceable use for a lot of old tin gutter troughs which were discarded during recent repairs in his house. The troughs, as illustrated, are cut into 2-ft. lengths, and into the middle of each piece a semi-circular sheet of tin is soldered. The irrigation guide when completed is embedded in the earth, with the central tin member directly in the center of the stream coming from the faucet. Obviously the flow of water is evenly divided and sent into the two trenches

By means of a division in the gutter-pipe, water is deflected in two directions

as desired. The earth beneath the hydrant is protected from erosion. Another great advantage of this simple device is that it permits alteration of the flow between the two trenches by merely moving the trough slightly in either direction.—JOHN EDWIN HOGG.

How to Prolong the Life of Flashlight Batteries

THE life of a flashlight battery can be increased if it is treated in the following manner. Before installing a new battery, take out the individual cells and give them a coat of sodium silicate, or water-glass, as it is commonly called. The entire cell should be coated, with the exception of the brass cap on the carbon and the bottoms. After drying, the battery may be reassembled and put into service. This treatment prevents the drying out that always shortens the life of dry cells.—THOMAS W. BENSON.

A NEW GAME - CATAPULTING

By Charles M. Miller

A NEW game has been devised, involving the old war tactics of catapulting. The device is known as a catapult. Instead of throwing stones or arrows, rubber balls are propelled distances ranging from 2 to 75 ft.; and, instead of city walls, a skeleton target lies flat on the ground. The target is usually placed by an opponent, and is located at an unknown range, but within the limits of the catapult.

The catapult has a moving lever A (Fig. 1) that is hinged to the base B and a strong spring C to draw it forward quickly. In the lever are cuplike holes in which a ball is placed after the catapult has been set, and when the lever is released it casts the ball forward.

If the lever has a full swing, it will cast the ball forward, but very little upward, as shown in Fig. 2; and, while the ball will be projected with considerable force, it will not go so far. If the lever is stopped part way forward, the inclination of the lever will determine the direction the ball will take, usually at right angles to the lever, or will leave it at this angle, which is shown in Fig. 3. If the lever is stopped far back, the ball will be flipped up in the air, but will make little progress forward, as will be seen in Fig. 4. The setting, therefore, that propels the ball for the greatest

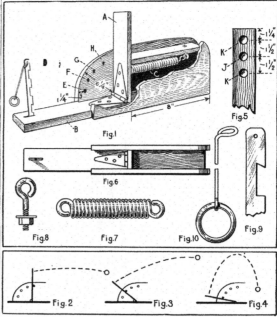

Details of the pieces for the making of the catapult and the assembled parts, with a diagram showing how the position of the lever affects the course of the ball. When the lever is vertical the ball shoots forward but not upward

distance will be with some elevation but not sufficient to waste energy in merely getting height. It is easy to understand why there are so many possibilities of range in the game.

There are many combinations that may be effected by the device. The lever may be drawn down to the last notch of the catch D (Fig. 1), or it may stop at two other points or notches, giving varying tensions as a starter. As mentioned, the lever may be allowed its full swing to the vertical position, where there is a permanent stop, or it may be stopped at several points by inserting a rod in the holes of the box. The holes are labeled E, F, G, H in Fig. 1.

Other Combinations

There is another series of combinations that may be used as distance modifiers. The three pockets shown in the lever (Fig. 5) have a varying effect from the longest to the shortest radius of the lever. The outer pocket, I, will cast the longest distance, because its arc is greatest, and the inner one the shortest distance, because its arc is smallest. These conditions make variations in the former combinations. For instance, if the lever is at full tension, the stop at F, and the ball placed at K, the cast will not be so great as it would be if the ball were placed at I; and so on.

Another variation of conditions is formed in the use of a hard and also hollow rubber ball. Three other conditions may be obtained by the elevation of the front end of the catapult by means of a brick. Raise it to the height of a brick flat, on edge and on end.

Much experimenting is possible with all these combinations, and it is a good study for a boy. If records are made of try-outs, considerable calculations may be made. In try-outs three casts should be made, the results added and divided by three for the data record. If an operation is faulty, it should not be counted. The try-outs should all be recorded in good form for future reference.

To Construct the Catapult

The base B, Fig. 1, is 24 in. long, $3\frac{5}{8}$ in. wide, and $1\frac{3}{4}$ in. thick. The sides are each 15 in. long, 8 in. wide, and $\frac{3}{4}$ in. thick. It is best to cut away the base for the length of the side pieces, as shown in Fig. 6, for appearance, and the front ends of the sides rounded at the top on a $1\frac{1}{2}$-in. radius and the back curve to correspond to the movement of the propelling lever, which has a radius of $6\frac{1}{4}$ in. The center of the back curve is on a line with the upper side of the base and at the bottom end of the lever.

The propelling lever A, Fig. 1, is 13 in. long, $2\frac{1}{4}$ in. wide, and $\frac{3}{4}$ in. thick and is hinged to the base with a large strap or door hinge less than 2 in. wide. The front face of the lever, when standing in a vertical position, is 8 in. behind the front end. Three pockets 1 in. in

diameter are cut in its front face. These pockets should be ½ in. deep. These may be bored with an ordinary bit, but it is better to use a spoon-bit. The center of the first pocket is 1¼ in. from the upper end, the second 1½ in. from the first, and the third 1½ in. from the second.

For the top of the box a piece of wood ¾ in. thick and as wide as the inside measurement of the box, for

The catapult with the arm drawn back and the ball in place for throwing it

this catapult 7 in. long and 2½ in. wide, is fastened in 1 in. down from the upper edges of the side pieces and 1 in. back from the front end. This makes the last stop for the propelling lever. It also carries the spring.

The spring is a very essential part of the device. It is a 1-in. closed spring 2½ in. long, not including the loops at the ends. Such a spring can be purchased from a hardware store. If it is difficult to obtain a spring just the right length, one may be made from a longer length by cutting it off and turning up a loop at each end. The spring is shown in Fig. 7. The attachment must be very secure. If eye-bolts, as shown in Fig. 8, are used, there will be no future trouble. These are not always available, and a large screw-eye may be used instead. The forward bolt or screw-eye is located 1 in. back from the front end of the upper piece in the box, and in the center lengthwise. The bolt or screw-eye in the lever is placed 3 in. from the lower end.

Method of Procedure

Since the spring loop will not reach the eyes of the bolt or screw, it must be stretched to hook in place. This is not easy to do. Make the attachment to the lever first, and then loop a strong piece of cord on the other hook and draw it through the box. Pull it to place, and with a screw-driver or other metal tool force the spring to one side, kinking it a little until the hook starts into the eye, then release the cord.

The sides of the box should be held together while boring the holes. This is done before they are nailed to the base. Be careful to bore the hole in straight, that is, at right angles to the surface of the board, boring through both in one operation. If the holes are ⅝ in. in diameter, a ⅝-in. pipe may be used for the stop. The pipe may be 6 in. long. The holes must be located so that they will be outside of the arc of the spring. To be sure of it, make the locations 1¼ in. from

the outside edge of the semi-circle on the side pieces.

The catch and release for the lever, when drawn down, must be simple and effective. The pull on the catch should be straight up from its hinge. A piece of soft steel 7 in. long, ½ in. wide, and 3/32 in. thick will be found very satisfactory. A hole large enough for an 8d common nail is drilled ½ in. from the lower end, and a similar one ½ in. down from the upper end. The notches are shown in Fig. 9. These can be filed into the metal, but a quicker way is to use a hack-saw. A wire and ring (Fig. 10) are used to pull the catch back for the release of the lever.

The catch-bar is inserted in a slot that may be made in the base with a compass saw after some holes have been drilled as starters. The holes must go all the way through, especially where the catch-bar is inserted. A hole is drilled in the side of the base to the slot where the hinge is located. To locate the catch-bar, draw the propelling lever down to the base, then draw a line across the base at the end of the lever, which locates the front edge of the catch-bar. This bar should stand in a vertical position when in use. There should be a small metal bar projecting ⅛ in. over the end of the propelling lever to make a good surface for the catch. The wood will not stand the wear.

When It Is Ready for Use

The device at this point is complete as far as casting the ball, but some careless person is apt to give the lever a hard knock if there are no guards provided. The guards shown extend to the end of the lever. They are made of flat iron. It will require

After the throw the arm stands in an upright position against the stop placed through the sides

about 6 ft. of ½ in. metal ⅛ in. thick. Two holes in the rear and one in the front end are drilled for screws, and to have the heads flush with the surface they should be countersunk. The whole device will present a better appearance if it is well painted.

The target may have an irregular form like the walls of a castle; or it may be made up of three laths laid end to end; or a folding target may be made. One person can obtain amusement, but it is more interesting to have an opponent whose business is to lay the target at an unknown distance from the catapult and return the balls to the operator.

Each turn consists of three trials. To make a score, the balls must hit the target, frame included, before striking the ground or other object. The first hit counts 10, the second 5. With another turn the target should be moved to a new location. The game consists of ten rounds for each side, not counting repeats. If more than one person plays on a side, they take turns.

A Combined Knife and Gouge for the Woodworker

WHILE at a Western summer hotel, one member of a party wanted some embroidery hoops, and

Fig.1

The end of the blade is turned over and sharpened so that the cut is made by drawing the knife

Cutting edge

Fig.2

an old Indian was induced to make the hoops. The hoops were made as ordered, but it was the tool the Indian used to make them that excited interest.

It was a combined knife and gouge, made out of a thin-bladed kitchen knife that had been curved at the end and then bent over and sharpened all along the edge and around the bend. The knife is shown in Fig. 1, and the bent end in Fig. 2. The way to use it is shown in Fig. 3. The depth of the cut is regulated by the angle at which the knife is held.

Timbermen use similar tools for marking lumber, and the Indian may

Fig. 3. How to draw the knife over the board to gouge out and make a depression

have taken the idea from them. But the little knife he used was a most convenient tool and not difficult to make.—TUDOR JENKS.

How to Make a Mat Surface on Aluminum

TO impart to aluminum the appearance of mat silver, this simple method will be found very satisfactory. After thoroughly cleaning the article, plunge it into a hot bath consisting of a ten per cent solution of caustic soda saturated with common salt. Permit it to remain in the prepared solution from 15 to 20 seconds, then wash and brush it. Place it in the solution again for half a minute, wash it, and dry it in sawdust.

A Combination Stock Barn Under One Roof

THE principal building on a farm may be a combined cow, horse, and calf barn, as illustrated. It is laid out in the form of a cross; not that this is the most economical shape to build, but it costs less than separate buildings. There are other good reasons for this shape. Having so much of his work under one roof saves much labor for the farmer. In some States the dairy regulations will not allow the young stock and horses to be in the same room with the cows. The cross-shape plan gives a three-in-one stable arrangement, bringing them together so far as the chores are concerned, but keeping them separate by merely closing the sliding doors.

Where a number of cows and several horses are kept, and there is young stock, this makes an ideal arrangement for housing them and for storing hay for feed. There are driveways each way through the barn, and yet none of the space is wasted. There are steel tracks to carry the ensilage and the manure, which may be put in the yard by means of the swing boom, or loaded on the wagon or spreader either in the barn or outside.

The mow floor plan shows how the hay is put down through chutes at points where it is most convenient for feeding all the stock. The large grain-bins have spouts in the corners near the wings, so that the grain may be delivered close to where it is fed. It is taken up through the door near the silos by using a rope and the steel track. The large mows have no timbers in the way of handling the hay, either when filling them or in taking the hay out, as this is a plank truss barn.

The large mow holds 110 tons and the smaller ones 40 tons each. By the covered passageway the center of the mows may be reached even when the large ones are full to the peak over the passage and grain-bins, as well as over the cows.

As there is only one way to enter the mow through the feed-room of one silo, the upper part may be locked by fastening one door. There is a water-tank in the horse stable, and water-buckets for the cows; and water may be piped to the stock pens.

The track in the horse stable runs into the harness-room, and a carrier can be arranged on which the heavy harness may be hung and all shoved back out of the way.—JOHN UPTON.

The cross-shape has its advantages for making a combination stock barn

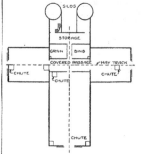

Plan of the mow for storing the hay and the location of the grain-bins

Renovating and Restoring Luster on Aluminum Ware

DISCOLORATIONS may be removed from aluminum ware and the original luster restored if the following directions are carried out. Wash the discolored pieces in a solution composed of 30 parts of borax dissolved in 1000 parts of water, adding a few drops of ammonia, and then drying them thoroughly.

A Simple Base for a Small Battery Lamp

AS it is often desirable to use small battery lamps when no base is procurable, a temporary base can be easily and quickly made.

A spool end is used for a miniature lamp base

Saw a common spool in half, and ream out the hole until it will thread in the base of the lamp. A small screw in the side of the spool serves as one terminal. The other consists of a spiral wire extending up in the bottom of the hole and attached to a screw in the opposite side of the spool. The construction of this small battery lamp can be readily seen in the illustration.—L. B. ROBBINS.

Sliding Resistances Made in a Solution with Zinc Plates

IT is essential in some experiments to have a varying resistance in the electrical circuit by a gradual, continuous process, and not by steps. There are two convenient forms of sliding resistances. One is a long resistance wire stretched backward and forward across a square wooden frame. Contact can be made at any point by a suitable sliding binding-post.

The other consists of a tall vertical cylinder of glass containing a concentrated solution of zinc sulphate, and having as electrodes at the top and bottom two zinc plates, one of which is movable up and down and can be held at any desired point by a suitable clamp.—PETER J. M. CLUTE.

Clips to Carry Policeman's Club on Motorcycle Handle-Bar

A NEW YORK motorcycle policeman has invented a very useful device for holding his club on the handle-bar of his machine. The de-

The clips on the handle-bar makes a convenient place for carrying the policeman's club

vice consists of two clamps fastened to the arms of the handle-bar. The club is easily slipped into the clips, and the spring in the metal holds it solidly in place.

The club is in a handy place, and can be taken up quickly whenever it is necessary to use the club.—PETER P. LEMBO.

Make Your Mind a File—Not a Pile—Stop Forgetting

By Prof. Henry Dickson

PROF. HENRY DICKSON, *America's foremost authority on Memory Training and Principal of the Dickson Memory School, Hearst Bldg., Chicago.*

IS your mind like a scrap pile—heaped up with a lot of unrelated, unclassified, unindexed facts? When you want to remember a name, place or date, must you grope uncertainly in this mixed-up pile seeking in vain to locate the desired information? And finally, in embarrassment, give it up? Summoned on any occasion to give facts and figures—does your mind become a blank? When suddenly called upon to speak—do you seek wildly to collect your thoughts—utter a few commonplace remarks—and sit down—humiliated? *Without Memory, all the knowledge in the world becomes worthless. "Stop Forgetting" makes your mind a file—not a pile.*

I CAN MAKE YOUR MIND AS SYSTEMATIC AND FORGET-PROOF AS A CARD INDEX FILE

The average mind resembles a scrap pile.

—master of your mind's infinite ramifications—instead of a victim of its disordered details. My course of Memory Training perfected by 20 years' experience, is universally recognized as the most thorough, practical and simplest system of its kind now before the public. My system so thoroughly trains the memory that you will be able to classify impressions, ideas, names, facts and arguments and have them ready at a moment's notice. It develops concentration—overcomes self-consciousness,

bashfulness—enables you to address an audience intelligently without notes.

IMPORTANT NOW

No time has been more opportune than the present to train the memory and the powers of concentration. The soldier needs this training to help him to master quickly the multitude of technical instructions which are part of his military discipline. The man who remains at home needs this training because business from now on will be more intensive than ever before. Every man in every place will find a reliable, efficient memory an asset of the utmost value. Whatever may be your position, send now for information.

DICKSON MEMORY TRAINING HAS HELPED THOUSANDS

Mail coupon or send postal for statements from students who had exceedingly poor memories and developed them to perfection—and men with remarkably good memories, who made them even better. Give me 10 minutes daily, and I will make your mind an infallible classified index, from which you can instantly select facts, figures, names, faces, arguments. *Perfect your memory and you can command what salary you will.*

The Dickson Trained mind is as well ordered as a cross-indexed file.

SPECIAL OFFER ON "HOW TO SPEAK IN PUBLIC"
This de luxe, handsomely illustrated, richly bound book—regularly priced at $2—free to every student who enrolls. The book will train you to think on your feet—to express yourself clearly, logically and convincingly, whether talking to one person or a thousand.

GET MY BOOK ON "HOW TO REMEMBER"
Simply send your name and address on the coupon or postal for this remarkable book. I will also send you a free copy of my unique copyrighted Memory Test.

Prof. Henry Dickson, Principal Dickson School of Memory
1929 Hearst Building, Chicago, Ill.

Send me your Free Book "How to Remember," also particulars how to obtain a free copy of Dickson's "How to Speak in Public," also Memory Test free.

NAME

STREET

CITY _____ STATE

How to Make a Half-Model Battleship

Plans and the way of cutting out the blocks to make a relief half-model battleship

By Joseph Brinker

HERE you are, boys. The illustrations will give you all the information you need to know how to make a half model of one of the most modern battleships. All that you will need are wood, glue, paint, some tools, and a desire to have a model that will be an excellent ornament for your den.

When you read of the patrol work Uncle Sam's great overseas fleet is performing in the North Sea, and of somewhat similar to those used in our navy. This is not an exact copy of the United States ships, since the plans of all navy vessels must be kept secret so that enemies may not obtain any information that will be of assistance to them. Nevertheless, the boat shown is modern in every respect, formed to give great speed, and long enough to give the necessary displacement for carrying the largest gun battery now

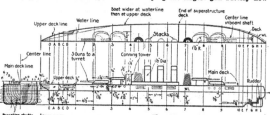

The section of the midship, showing how to lay out the different curves from stem to stern by the letters and figures representing the cross-section location on the elevation view

perhaps some big sea battle which our sailors and those of the Allied navies staged sometime before the Kaiser acknowledged that he was beaten for good, think how proud you will be if you can show your friends a fine model of the type of super-dreadnaught which played the most important part in the fight.

The illustrations show a super-dreadnaught of the battle-cruiser type,

afloat, twelve 14- to 16-in. guns, three each in four turrets located on the ship's center-line, two forward and two aft.

The position of the basket-masks, the conning-tower, and the smoke-stacks are given, together with the position of the secondary battery below the main-deck and the contour of a typical bow and stern, with the location of two of the four propel-

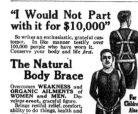
Holding block

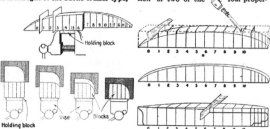

Vise   Blocks

Holding block

Cardboard cut from the cross-sections, as shown on the next page, are used to shape the block of wood; indicating the method of sawing and cutting waste material

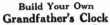

lers which drive the vessel at a speed of close to twenty-five knots an hour.

In general, the model is 56 in. long over all, about 4¼ in. wide, and about 10 in. high from the bottom of the keel

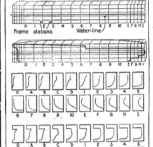

Frame stations   Water-line

Frame C

This edge must be straight

The black of wood as it is marked with location figures and the cardboard forms

to the top of the smokestacks. It gives a true form of half of the ship, and should be mounted on a painted backboard about 60 in. long and 12 in.

Manner of gluing several pieces of wood together to make a large block for the model

high. This size enables the maker to embellish the boat with miniature guns, anchors, flags, and even lifeboats if desired.

The model is made of pine, since that is the cheapest and easiest to work

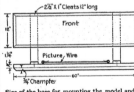

2½" X 1" Cleats 12" long
Front
12"
1¼"
Picture Wire
60"
¾" Champfer

Size of the base for mounting the model and the way the wires are attached for hanging it

with. It is glued together as indicated to form a solid block, out of which the shape is cut with a chisel, and then smoothed with sandpaper.

To be made realistic, the portion under the water-line may be painted a bright red and all above the water-line war gray.

FOR THE MAN WHO CARRIES TOOLS

This Handsome, Substantial Harness-Leather TOOL BAG

Index for Fluting Columns on a Shaping Machine

THE fluting or beading of a column or a table leg is apparently a simple piece of work, but it requires a device like this to do it economically and accurately. Work of this sort is usually done on a shaper, and an index similar to the one illustrated is a necessity in a shop that does miscellaneous work.

The base A may be made in any

An index head for mounting on a fluting machine to space the flutes on the circumference

length, but 6 ft. will allow a column 4½ ft. long to be handled between centers. It should be made of straight, well seasoned wood cut away at B to permit the shaper head to run close to the edge of the base. The head of the index C consists of a threaded block mortised into the base and pinned as shown at D. It is plain that the lower the center of the column and the lower the cutter is set on the head, the less vibration there will be on the work and the results will be more satisfactory; but the capacity of the index will be greater if there are two centers instead of one, so that columns of different diameters can be fluted.

The screw at E should be not less than 1 in. in diameter, or it may not be rigid if it projects more than 2 in. from the head to reach the column at F. A stout spur center at G may be made by turning in a screw and filing it to a point. If the screw E is loose enough to turn easily, it may turn back as the work proceeds; but this may be prevented by boring a hole at H and driving in a wedge-shaped pin, so that it will bear on the screw.

The essential mechanism of the device is the traveler J, which slides in the groove K of the base and is adjustable to different lengths. It is held in place with pins at L. The spur center M must exactly coincide with that of the head. The indispensable part of the index is the plate N, which consists of a piece of hard wood, or, better still, a piece of veneer. The edges must be scored with slots spaced accurately on the outside as at O. A plate may be made for each number of cuts desired, regardless of the diameter of the column, but one may be used for different spacings. For

instance, a plate with twenty-four scores or saw kerfs in its edge will index any column that requires in its circumference 2, 3, 4, 6, 8, 12, or 24 flutes. The square hole in the center of the plate fits closely on the square end Q of the plug R, which carries the index plate and turns in the hole S in the traveler block as the column is turned. It may be made to work easily by waxing it. The spur center M of the plug and the spurs T on the face of the plate should be made of heavy screws turned in, and the center spur M must be filed to a sharp point and the spurs T filed somewhat like an awl point.

The metal washer at Y will permit the index plate to turn easily by reducing friction, which causes the plate to jump when it is moved forward for the next cut. The back of the index plate, the face of the traveler block and the washer may also be waxed to reduce friction. When the column is set in place between the screw center G and the plate center M, it and the plate may be turned by hand.

The piece U may be screwed on the side of the traveler flush with the face, and scored at a point that coincides horizontally with the center of the plug M. The scores O in the edge of the index plate and the scores V in the piece U should be made with the same saw to insure a uniform width to receive the feather W, which acts as a lock to hold the plate rigidly while the knives are cutting. The feather W should be filed slightly wedge-shaped to fit the scores closely. By turning the index plate N until a score in its edge coincides with score V and inserting the feather W, the column will be held firmly while the shaper is doing its work. When the flutes in the column are cut, they will be as accurately placed as are the scores on the edge of the index plate, and any inac-

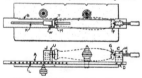

How the work is mounted on the machine and in the index head for fluting

curacy in the plate scores will be increased or decreased as the column is larger or smaller than the index plate.

The bottom of the tenon of the traveler V should be saw-kerfed as at X, and these filled lengthwise with a tongue, glued to prevent the wood from pulling out by the draw of the pins L. The holes in the tenon to receive the pins L should be bored slightly higher than the holes in the base, to draw the shoulder of the traveler into close contact with the top of the index.—CHARLES A. KING.

These Are The Hours That Count

MOST of your time is mortgaged to work, meals and sleep. But the hours after supper are *yours*, and your whole future depends on how you spend them. You can fritter them away on profitless pleasure, or you can make those hours bring you position, money, power, *real success* in life.

In these days, when men are being called to the colors from office, store, shop and factory, important places must be filled quickly and the boss wants trained men to fill them. You men of middle age, you who are too young, you who for one reason or another cannot go, this is *your* chance to serve your employer and your country and at the same time win promotion and a salary that will mean more comforts and happiness for those dependent on you.

There's a big job waiting for *you*—in your present work, or any line you choose. Get ready for it! You can do it without losing a minute from work, or a wink of sleep, without hurrying a single meal, and with plenty of time left for recreation. You can do it in one hour after supper each night, right at home, through the International Correspondence Schools.

Yes—You Can Win Success in an Hour a Day

Hundreds of thousands have proved it. The designer of the Packard "Twin-Six" and hundreds of other Engineers climbed to success through I. C. S. help. The builder of the great Equitable Building, and hundreds of Architects and Contractors won their way to the top through I. C. S. spare-time study. Many of this country's foremost Advertising and Sales Managers prepared for their present positions in spare hours under I. C. S. instruction.

For 27 years men in offices, stores, shops, factories, mines, railroads, in the Army and Navy—in every line of technical and commercial work—have been winning promotion and increased salaries through the I. C. S. Over 100,000 men are getting ready *right now* in the I. C. S. way for the bigger jobs ahead.

Your Chance Is Here!

No matter where you live, the I. C. S. will come to *you*. No matter what your handicaps, or how small your means, we have a plan to meet your circumstances. No matter how limited your previous education, the simply written, wonderfully illustrated I. C. S. textbooks make it easy to learn. No matter what career you may choose, some one of the 280 I. C. S. Courses will surely suit your needs.

Make Your Start Now!

When everything has been made easy for you—when one hour a day spent with the I. C. S. in the quiet of your own home will bring you a bigger income, more comforts, more pleasures, all that success means—can you afford to let another single priceless hour of spare time go to waste? Make your start right now! This is all we ask: Without cost, without obligating yourself in any way, put it up to us to prove how we can help you. Just mark and mail this coupon.

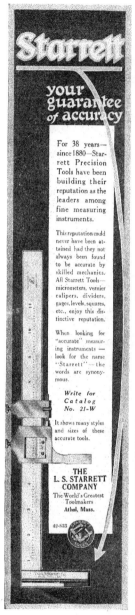

Removing the metal base from the globe without breaking the glass

Filing through the crown holding the glass rod and its filaments

Making Vases Out of Electric Light Bulbs

VASES that are both decorative and useful may be made from burned-out electric-light bulbs. Take a burned-out bulb and hold the metal part of it in a flame. After a few moments turn the metal screw cap with pliers until it comes off. Then break the wires holding the two parts together. File through the top crown holding the glass rod and its filaments. Then break off the crown and take out the rod.

Make a stand for the bulb out of brass or iron strips 1-16 in. thick and 1-4 in. wide. This can be done by bending the strips into any desired shape, such as the one shown here, with a pair of flat-nosed pliers.—E. BADE.

An attractive vase filled with flowers

A mounting made of metal strips

How to Construct a Sponge-Box or Bread-Raiser

A SPONGE-BOX or bread-raiser is a temperature equalizer for the housekeeper who finds the fluctuating heat and cold waves an aggravating circumstance in making bread. Hold-ing the sponge or dough at the proper temperature lessens the time required for the substance to rise.

Such a box can be made from a dry goods packing-box. A convenient size is 20 in. square and 26 in. high. About 10 in. from the bottom of the box is placed a shelf made of slats and strips of wood on cleats fastened to the sides of the box. A second shelf is placed 4 in. above the lower one. The shelves can be removed when cleaning the box. Below the lower shelf a sheet of galvanized iron slightly wider than the shelf is inserted. It is curved in order to make it slip in and stay securely. This prevents scorching of the lower shelf when a lamp is placed below, and it also helps to distribute the heat evenly. The door is hinged, and fastened with a thumb-latch or a hook and staple.

A number of small holes are bored in the lower and upper parts of the sides and in the top of the box to promote the circulation of air. A cork, which has been bored through the center to admit a straight thermometer, is inserted in one of the holes in the top of the box. A Fahrenheit chemical thermometer that registers as high as 100 deg. can be used.

To avoid danger of fire, the box

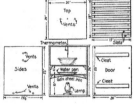

The interior of the bread-raiser, showing the location of the shelves and the lamp, also the working plans with details for making the box at home

should be lined with asbestos or tin when a kerosene lamp is used for heating the box. If an electric light is used, the lining is unnecessary. A 16 c.p. light will heat the box amply. A small and inexpensive night-lamp is placed in the box, and a shallow pan of water is placed on the lower shelf, so that the air in the box will be kept moist.

The bowl of sponge or pans of dough are placed on the upper shelf. The temperature of the box should be kept as near 86 deg. as possible when the bread is being made in the quick way. If a sponge is set overnight, the temperature should be kept at 65 to 70 deg. until the dough is made in the morning, after which the temperature may be increased to 86 deg. The temperature in the box may be varied by raising or lowering the flame of the lamp, or by changing the temperature of the water in the shallow pan. —S. R. WINTERS.

How to Keep a Miter-Box from Warping

SHRINKAGE of the sides of a miter-box does not affect its accuracy, but the shrinkage of the bottom will throw out of line the cits in the sides, and render the box worthless for accurate work. The illustration shows a simple and infallible means of preventing, or rather counteracting, this shrinkage. It consists, in effect, in splitting the miter-box longitudinally through the bottom and connecting the halves by transverse pins fixed to the box only at the extremities.

In practice this result is accomplished by using two pieces of heavy

A split base with cross-pins to prevent the miter-box from warping

material for the bottom, the combined width of which should be ⅛ in. less than the desired width of the box. These pieces should be carefully jointed and squared all around and the side pieces nailed on. Place the halves together on the work-bench, adjust them carefully, and clamp them firmly together. Bore a ¾- or ⅞-in. hole through the sides and the bottom at each end of the box. Insert a dowel, which should give a very light drive fit, in each hole, and drive a nail through each end of each dowel.

This arrangement holds the sides of the miter-box in the same relative position, and permits any amount of lateral movement of the bottom under changing atmospheric conditions.

If good material is used for the bottom and a good job done, it will pay to go farther by screwing the sides on instead of nailing them, thus making them easily removable when they become worn.—HENRY SIMON.

A Simple Type of Home-Made Magnetograph

IT is known that there are continuous changes in the direction and strength of the earth's magnetic field, such variations being at times so marked as to merit the title of "magnetic storms." The origin of the

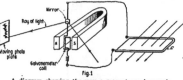

Fig.1

A diagram showing the way a magnetograph records the strength and direction of the earth's magnetic field

variations noted at any observation station may be remote or purely local. They may arise from solar disturbances, volcanic eruptions, or even from local climatic conditions, such as cloud shadows or showers or wind gusts, as recently shown by Nipher.

For many years progressive governments have maintained magnetic stations. The observations at these stations are made by suspended magnets bearing small mirrors, which make continuous records of their ever-changing angular positions by the reflections of the beam of light on a moving sensitized photographic film. The path of the wandering spot of light on the film is permanently revealed by the subsequent developing and fixing process.

More than thirty years ago another method of registration of magnetic disturbances was put into limited practice. This was the registration of the deflections of a sensitive electrical current indicator (a galvanometer) connected to a simple coil of wire (Fig. 1). Variation in the number of lines of force of the earth's

Fig. 2. A record made with the coil lying flat on a horizontal plane

magnetic field which pass through the coil *C* produces a current through the galvanometer coil proportionate to the time rate of that variation. Hanging between the poles of a steel magnet, *N-S*, the galvanometer coil deflects in proportion to the magnitude of the induced current traversing it.

Some of the lately developed types of low-resistance, high-sensitivity galvanometers yield extraordinarily fine

magnetograms. In Figs. 2 and 3 are shown two records recently taken at Washington University, using an American-made galvanometer in connection with a circular coil 1 ft. in diameter and containing 800 turns of wire. A narrow beam of light from a 100-watt mazda lamp was reflected from the galvanometer mirror to a photographic plate undergoing a steady lateral shift of 1 in. in ten seconds.

In Fig. 2 is shown a record with the coil *C* laid flat upon a horizontal table-top, and Fig. 3 shows the record with the coil standing on edge, with its plane approximately north and south—a position in which the coil embraces comparatively few of the lines of the earth's magnetic field.

The more violent disturbances indicated in the records are due to the proximity of an electric car line 600

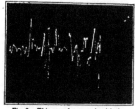

Fig. 3. This record was made with the coil on edge pointing north and south

ft. distant. Certain U. S. magnetic survey records show undoubted disturbances arising from an electric car line at a distance of thirteen miles from the observing station.

It has been suggested that a coil such as the one shown in *C*, Fig. 1, could be laid upon a harbor bottom, in connection with a shore galvanometer, would indicate magnetic disturbances arising from the steel hull of a passing submarine. But, considering the ever-present normal fluctuations, grave doubt may arise as to the efficacy of such a device.—LINDLEY PYLE.

How to Put Packing in an Automobile Pump

IN packing the circulating pump of an automobile the following method will improve the work. Determine in what direction the shaft runs; then, when winding the strand of packing, wind it in a direction by which the friction of the shaft will tighten up the spiral rather than loosen it, so that the packing will be in a slight measure self-tightening—the reason being the same as for the use of right- and left-hand threads on the parts that revolve and are subject to friction against stationary members. A pump so packed will give very little trouble.

the development of the extremely high-speed motor the weight of the pistons was reduced to a minimum. In many cases these light-weight pistons have given trouble, and usually it is because of the lack of care in perfecting the design.

A groove cut connecting pin with ring slot

In one case a well known car with aluminum pistons developed a knock after only a few hundred miles of service. The cause proved to be a loose piston pin, and the matter was remedied by removing the piston and fitting a slightly larger pin. But soon the trouble showed in another cylinder. This time all the pistons were altered as shown in the illustration, and no further trouble was experienced.

The lower piston ring was removed, and the upper corner of the lower side of the groove was filed at an angle of 45 deg. for a short distance on each side of the pin, as at *A*; then a small groove was cut connecting this to the pin bearing. The groove below the ring was cut to catch the oil, and it was then drained to the pin bearing, which evidently had not previously received sufficient oil.—S. E. GIBBS.

Use Solder to Secure an Iron Pin in Wood

I HAVE found that solder is the quickest thing with which to fasten a long pin in a section of hard wood (Fig. 1), glue requiring too much time to set and also preventing subsequent removal. After tinning the pin I ran plenty of solder around it (Fig. 2) just where it emerged from the wood. The

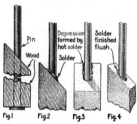

Fig.1   Fig.2   Fig.3   Fig.4

The different stages in making a connection of a metal pin in wood

Outwit Old Man Winter

DECIDE now to protect your car during the coming winter with Johnson's Freeze-Proof. Purchase your supply early from your dealer and read and follow the directions carefully. A little time spent now in cleaning the radiator and putting on new hose connections will save you unlimited time, trouble, worry and expense during the winter months.

JOHNSON'S

·is the logical anti-freeze preparation. It is inexpensive—does not evaporate—is non-inflammable—easy to use—and guaranteed. One application will last all winter unless the solution is lost through the overflow pipe or leakage.

Truck and fleet owners will find Johnson's Freeze-Proof a great time and money saver. Your trucks will always be on the job and in the coldest weather it will be "Business as Usual" for you.

Farmers will find Johnson's Freeze-Proof a utility product—for automobiles —tractors—gas engines — trucks — and electro lighting and heating plants.

S. C. JOHNSON & SON,

The present high price of alcohol —its low boiling point—quick evaporation and inflammability make it impractical. Use Johnson's Freeze-Proof, then forget there is such a thing as a frozen radiator.

One package will protect a Ford to 5° below zero and two packages will protect it to 50° below zero. See scale on package.

Racine, Wis., U. S. A.

All the Year 'Round

INDOORS and OUTDOORS

Ease the throat with—

LUDEN'S MENTHOL COUGH DROPS

The many uses for Luden's have won year 'round popularity. Luden's quickly relieve huskiness, throat tickle, dryness and similar irritations. No matter who you are or what you do, you'll find Luden's helpful and handy.

Look for the familiar Luden yellow package

Made by
WM. H. LUDEN
in Reading, Pa.
Since 1881

GIVE QUICK RELIEF

"My children chanced to give me a stick of your shaving soap," says Mr. Jones. If you do not have one given you—give yourself a "Handy Grip."

"In the directions was an admonition not to rub...I thought this extremely silly," says Mr. Jones. But he tried it and found the truth of the original Colgate phrase used since 1897—"needs no mussy rubbing in with the fingers."

"I am, from now on, a firm champion of Colgate's Shaving Stick. It beats anything I have ever used," says Mr. Jones. You will reach the same conclusion after you have tried Colgate's "Handy Grip."

Colgate & Co.,   New York, April 6, 1918.
New York City.

Gentlemen:

Some months ago my children chanced to give me a stick of your shaving soap, and I am writing to you to tell you what a delight shaving now is with it, instead of the nightmare I formerly experienced whenever I shaved. The lather is wonderful, smooth as velvet, leaves the skin without any irritation, and in short, your shaving stick is ideal for the purpose.

In the directions that came with the shaving stick was an admonition not to rub the beard with the hand after applying the lather. I thought this extremely silly, but on trying it I found that your experts had solved another important element in successful shaving and I shall never again do anything but follow the advice given. The brush, applying the lather with the proper degree of moisture in it, is sufficient. The new method is vastly superior to the old.

I am, from now on, a firm champion of Colgate's Shaving Stick. It beats anything I have ever used, and I've been shaving myself for 35 years.

None of your agents have solicited this testimonial from me but I send it merely as a deserved recognition of the makers of the best shaving soap in the world.

Very truly yours,
(Signed) EDWIN J. JONES*

*Mr. Jones is Associate Editor of The Financial World

THE "HANDY GRIP" is the thrifty Shaving Stick. It saves you 50 shaves below the "Waste Line," and more: you can buy a Refill Stick for the original metal Grip.

CPSIA information can be obtained
at www.ICGtesting.com
Printed in the USA
BVHW04*1209180918
527831BV00013B/853/P

9 780484 414340

Lun

By Xanthe E. Horner

Erebus Society

Erebus Society

First published in Great Britain in 2021
by Erebus Society

1st Edition

Copyright of Text Placement © Xanthe E. Horner 2021
Cover & illustration copyright © Xanthe E. Horner 2021
Editor: Constantin Vaughn

ISBN: 978-1-912461-27-1

www.ErebusSociety.com

Lunations

Ancient pull
of dark and light
Ignite the shadows
Born unto night
Guide us Nuit
Into the abyss
The depth we plunge
The height of flight

Contents

After & Before

I had a dream about you
Rearranged the coordinates
The distinction between touch and absence

I felt you on the underside
Of that vast, shining disk

The knowledge of it
Hanging apparition in the sky

Great black glimmering void
You have held us
In the absence
In the tension between fusion and null

Death was heard in the hiss of a mouth
Opening,
An electric impulse
That conjured the spectral image of itself
Its last motion, an expulsion
The meaning for itself

The reason and act
The oceanic crackle of a wave's break
Played on an analogue wire

Half-formed, I'd seen
A skeletal imprint
The black dog

Great black glimmering void

You aren't
I am
I am not that you are

The searchless spot
In the apprehension of light

Origin of quintessence
The death of stars
The birth of aeons

Faceless mage

The lover under the skin
The vacuum cling
Before dust

Selene

All of the things inside
The treasure and rot
She sifts
Under moon's tide,
Reminiscing what's not and is
Under night like this

Selene, a pale face
Mark of serenity
The spit-shined pearl
Returned herself to a former glory

Holds herself up to the dark
She retains all that light
Traces on skin that she thought was begot
Now a blank immaculacy
Craters
Where the indentations of loveless passengers
Bruised

All trying to get somewhere
She convinced herself
She had to be a vehicle

Selene
Receptacle
Collects the dust from moody gloom
And shaking trees
That she blows in her palm
Scattered charm spun

The milk stained sky
Sees and knows all
Her mimicry chimes
Flattery

If she can be that interplay
Morse code fragmented
Twinkle-wink the pause
Between light and light

Dot dot and ignite
A lonely assurance
What isn't forever will break
Against your bones

Wield hands and heart
And gasp
It's passed, it's past

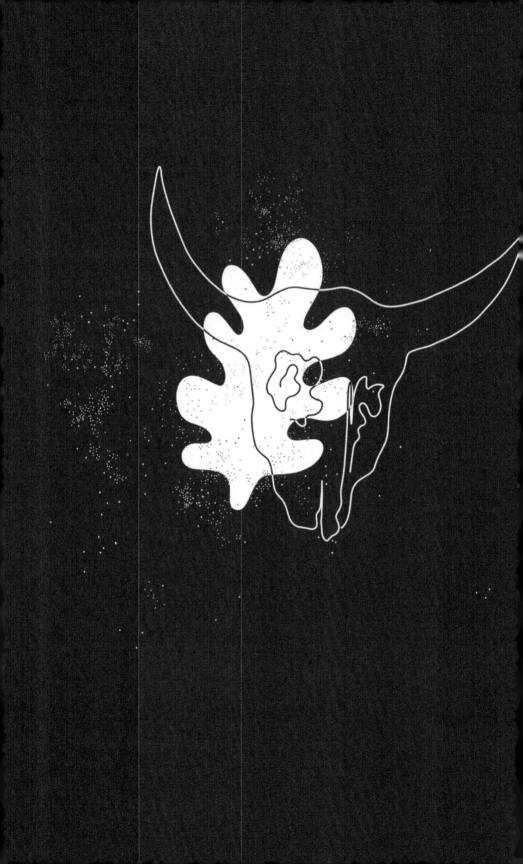

Hathor

They called me a heifer
Well, Holy cow
I've spread my ancient lunation
Starry lactation
I've split the pip of the dawn
Uttered the sacred vow

Call me a heifer
A dark priestess
Lunatic or lunarian
I lay my case to rest
Upright casket, scorpion neck
Cavity

That draws forth the vowel
Vibratory speech
That stings you

Suckle udder
Under slim sinew moon
Like sickle
Drip and trickle
Swell and boon

You called me a heifer
Well Holy cow

Tonight I'll undo
The shame of the human
Tread with me
Graze on silken meadow grain
Temporal haze
Animal stunned

A last palpitation
Before the clock is set
The hands of the sky beckon
The roaming fates

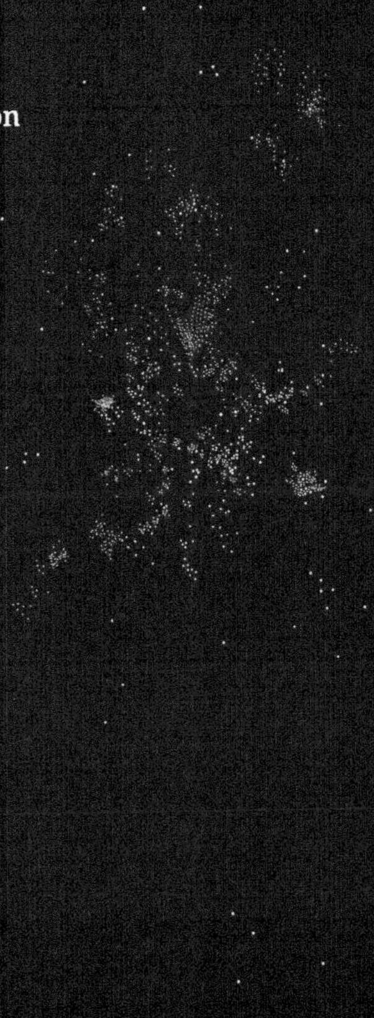

Ishtar

Enfolded in her naval
The glitter grit of time
Compressed interstellar sweat
Galactic pulse
A white, hot mouth
That grins
Ear to ear
Tells me she is from a time not here

Ishtar
Her name is an invocation
She announces herself
Issssh-tar
The drop of a flaming coal in a salt lake

Her naval the chalice
I lap
Hole to water human cattle
 As I feel her within
Churning up my insides
Retained moon tide

Swollen and bruised
Took a bite from ancient fruit
Exploded my delicate sensibilities

In another shade she is Inanna
The Holy lap
I kiss between toes
Initiate

On the tip of the tongue
Is an imagined future
Narrated by a past unseen
Queen of prehistory
The old come new

Captive hearts
Courted by deep saline mystery
That prises open
The forbidden box
Behind the mind

The reptile awareness
That carries itself on the atemporal
Shed skin for scales
Across salt and shale and sands

Ishtar-Inanna
The serpent bites
The willing hand

Hecate

Hellcat
With three faces
The infernal graces
She gave birth to herself
And herself

In all phases
Woman, refracted
Plain of knowledge
Cosmic immediacy
The swelling undercarriage of
The twilight

Ready to engulf the light
She is the lickety-split saliva
On a candle's wick
Ouch

Smoke winds in
The gates of the eternal
Opening
I wade in dark and darker
Mother

Woman in scarlet
She appears in your room
At the edge of your despair
Teasing back blankets

Her rapturous chanting
Fills ears
And hips thrust
She tells you to abide to all this is

My cloak is her vortex
A whirlpool at the end of the bed
Led by a dark hand
With pearlescent finger tips

Hecate
She spins
I'm drawn a shade paler
Outlined in sanguine crayon

She tells me this blood is my boundary
It keeps me alive and apart
Held together
In-difference
Armoured in the night

At 3:00 on the dot
She changes her face
Baying dogs
Violent moon
Metallic taste

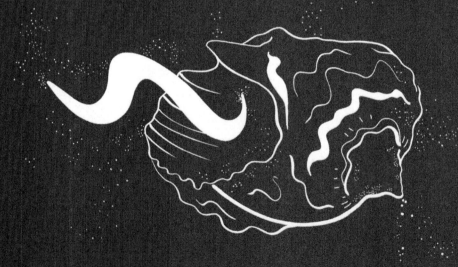

Sybilla

Scribe
From the foam of the stars
Etch wretched prophesy
Imbibed
Like long nights out
wasting and gazing

He takes his hand in hers
Buoyed up on her impermanence

A honeyed mouth that slathers
Slackens and aches before vowels
And Oh!
Sweet trembling wake
Ecstatic nonsensical quake
First the tongue
The waters sprung forth
Overcome

Nestled in her mystery
A finger edges closer to her moon
She is asphalt and pitch
The creeping dim that holds
The sliver of light
To pass between lips

Liquid cooling in a vessel
Wanting to cascade over rim
Such mortal whims set alight
The phantasms of delight
In those who seek to sip
To spill and soak the chin

Anchored to the fickle seas
The soft tug of rolling tide
The thought-numbing breeze
Sets and unsets her cheeks
Makes her cool to his need

All those flagrant affairs
Over in a week
Nuit guides her
Sybilla

Not made for mortal fixity
She seeks her own element
Nectar of divinity
Runneth over and over
The cup
She wants to be spent, supped
But she is boundless
Knowable only by night

Blanketed deity
Not a representation
But night, stark and black
More real than mortal caress
She holds, is held

The High Priestess

Collected musings of a sage
Condensed and weighed
Scent signature of night-blooming flower
The discharge from several moons
Powdered to dust
Stored In alabaster

Held up to show her mercurial power
Worth her weight in that strange pink fruit
That mimics the human heart
The juices spill over, pleasurable
Like the fox, or *fuchs*

Tall as the pillars she sits between
Monochromatically spliced
The polar twins of a magnet
Sucked
Thumbing the scroll
That turns over answers
If only the inner eye perceives
Saucer-wide to beckoning tide
Coquettishly pooled at her ankles

Modest habit she glowers beneath
Upturned, triple threat at threshold
Beyond which are the broken banks
Flooded abysm of wisdom
The apple and serpent are paired
For life and beyond
They flee on the Night Bark
The arcane tips into the known
Reservoir of fallen stars

She chants
Yod-Heh-Vau-Heh

Casts a net over the formless
The promise that soon
The ageless ache will soothe
Wriggle loose the milk tooth

Her hands form a prism
To receive and transmit
Transmute holy patterns left
By waves of the Styx
Dish and spoon
Brazen moon
As above so below her

She draws up from her feet
What abides and what's been
That moon will preside
Beside her
Hand and nail
Hook and Flail

Flower of the firmament
Elemental Daughter

The Empress

All is abundant here

She chimes, emphatic feline eyes
Teasing not a laugh
But a purr
Of fur-lined resonance
Full with child
Wed not to the plough
But the field
The ripening bough

Flowing dress spun from the silk
Of those small, luck-giving spiders
Ready to grant a wish
If only you cease to fear that
Which you can hold in both hands

The transmitted potency
Of her sister's lunations
Prised open to meet the dawn
A recurring bliss of golden hue
That greets as a kiss
Across sun-drenched face
Untamed locks framed by
Shooting Stellar crown
A mane

She sits upon her bounty
Mother to golden goose
Glows with the inevitability
Of change as lifeforce
Honeysuckle waters break
And fecundate the land
She basks and sighs

It shall be, of course

Hand clasped to sphere
A pledge to make gold the earth
Not Midas, but alchemist
Salve et coagula
Harvest of worlds
A cornucopia in a nutshell

Know thy worth

0

At the crossroads
Between two tracks
A vixen in half
Only torso

An upturned card
Three of spades
On its back
Reversed
But the sun is shining

The last waif that wound up
Dust
On my watch
Was really a beginning
Atrophy as the resultant force
Of heat
Entropic growth

Of bodies baying
In the spring night
and sharing disease
The full worm moon
Itching into bloom

The first one
I took her to the lake
Laid out on moss bank
Triangulated points of light
Kept her safe
Mallard feather
A Totemic wreath
For her journey
On the night raft

Our mother of mercy
Took her away
The bluebottles did what
They were made for
Life and life and life

The Trembling wake
Of another cycle's completion
This one has lost her head
Portends an end
Or was it about remission

The spades don't sting like they used to
My words are not petty cuts
But bandages for wounds

Healing hush and crush up leaves
For salve
I pour into earth
The hot libation
By the tree that you only find
 When you stop

Crossing my path
Is that fated orange mate
With singed paws
 As if dipped into flame
I remember
That it's not about being saved
Or trying to cheat fate

But when you meet her,
Lower your gaze

41

Printed in Great Britain
by Amazon